MW00785391

Acrocanthosaurus
Inside and Out

Acrocanthosaurus
Inside and Out

Kenneth Carpenter

UNIVERSITY OF OKLAHOMA PRESS : NORMAN

Also by Kenneth Carpenter

(coeditor) *Dinosaur Systematics: Approaches and Perspectives* (Cambridge, N.Y., 1990)
(coeditor) *Dinosaur Eggs and Babies* (Cambridge, N.Y., 1994)
Eggs, Nests, and Baby Dinosaurs: A Look at Dinosaur Reproduction (Bloomington, Ind., 1999)
(editor) *The Armored Dinosaurs* (Bloomington, Ind., 2001)
(coeditor) *Mesozoic Vertebrate Life* (Bloomington, Ind., 2001)
(editor) *The Carnivorous Dinosaurs* (Bloomington, Ind., 2005)
(editor) *Horns and Beaks: Ceratopsian and Ornithopod Dinosaurs* (Bloomington, Ind., 2007)
(coeditor) *Tyrannosaurus rex, the Tyrant King* (Bloomington, Ind., 2008)
(coeditor) *Dinosaurs and Other Reptiles from the Mesozoic of Mexico* (Bloomington, Ind., 2014)

This book is published with the generous assistance of the Museum of the Red River, Idabel, Oklahoma.

This book is published with the generous assistance of the Wallace C. Thompson Endowment Fund, University of Oklahoma Foundation.

Library of Congress Cataloging-in-Publication Data

Names: Carpenter, Kenneth, 1949–
Title: Acrocanthosaurus inside and out / Kenneth Carpenter.
Description: Norman : University of Oklahoma Press, [2016] | Includes bibliographical references and index.
Identifiers: LCCN 2016004770 | ISBN 978-0-8061-5393-3 (hardcover : alk. paper)
Subjects: LCSH: Dinosaurs—United States. | Carnivorous animals, Fossil—United States.
Classification: LCC QE861.4 .C37 2016 | DDC 567.912—dc23
LC record available at http://lccn.loc.gov/2016004770
The paper in this book meets the guidelines for permanence and durability of the Committee on Production Guidelines for Book Longevity of the Council on Library Resources, Inc. ∞

Copyright © 2016 by the University of Oklahoma Press, Norman, Publishing Division of the University. Manufactured in the U.S.A.

All rights reserved. No part of this publication may be reproduced, stored in a retrieval system, or transmitted, in any form or by any means, electronic, mechanical, photocopying, recording, or otherwise—except as permitted under Section 107 or 108 of the United States Copyright Act—without the prior written permission of the University of Oklahoma Press. To request permission to reproduce selections from this book, write to Permissions, University of Oklahoma Press, 2800 Venture Drive, Norman, OK 73069, or email rights.oupress@ou.edu.

1 2 3 4 5 6 7 8 9 10

To the many Works Progress Administration workers
who excavated and prepared the first *Acrocanthosaurus*,
thereby making this fantastic dinosaur known to the world;

to Cephis Hall (1941–2013) and Sid Love (1924–1997)
in recognition of their hard work excavating Fran,
the *Acrocanthosaurus* of McCurtain County,
which I was privileged to describe;

and to those schoolchildren who raised $150,000
to bring a cast of Fran to the
Museum of the Red River in Idabel, Oklahoma

Contents

Preface

This book is about *Acrocanthosaurus*, a *Tyrannosaurus*-size carnivorous dinosaur that lived 110–115 million years ago in what is now the south-central United States. I could write that it "terrorized" the region, but that would probably be an exaggeration. *Acrocanthosaurus* had a role in its ecosystem, just as the lion does in East Africa today. Zebras and gazelles do not necessarily run at the first sight of a lioness (the females do most of the hunting), but they do watch warily for behavior that shows she is hunting. It seems possible that potential prey kept a wary eye on a passing *Acrocanthosaurus* and fled only if it seemed to be on the hunt.

Acrocanthosaurus (ack-row-can-though-soar-us), or "Acro" for short, is known from four partial skeletons and a handful of bones from a fifth. Together they represent nearly all the bones of a skeleton, so we have a fairly good idea of what this dinosaur looked like. Studies of the joints show their range of possible motion, whereas bumps, ridges, and other scars on bones show where muscles, ligaments, and tendons attached. The bones have been studied using a variety of techniques, including X-ray computed tomography (CT scans) to peer into the braincase and microscopy to look closely at the insides of bones. These and other techniques give us a fairly good understanding of this dinosaur's biology.

Much of this information exists only in the scientific literature, which is listed in the references along with a brief explanation about the key point of each source. Some of these works, or at least their summaries (abstracts), can be found for free on the Internet with a Google Scholar (http://scholar.google.com) search; understanding them is another matter. This book attempts to distill this technical literature and other information to paint a picture of a truly remarkable dinosaur. To do this, I draw upon my knowledge of Acro from having coauthored a scientific publication with Canadian dinosaur paleontologist Philip J. Currie on the most complete Acro specimen, known by the nickname of Fran (Ferrell 2011).

I first learned of Fran in 1986, when I briefly worked for the Oklahoma Museum of Natural History, as it was called then. I did not know I was looking at the bones of Fran at the time, nor did I know then that I would eventually play a role with Fran by describing it.

Dr. Rich Cifelli, the paleontologist at the museum, showed me a few bone fragments that local resident Kristi Silvey from McCurtain County had collected around 1982 and turned over to the museum a few years later. When Rich visited the site in early April 1987, it was clear that someone had already been digging a large hole. Seven years after I first saw the fragments, I was working for the Denver Museum of Natural History when rumors reached me that an *Acrocanthosaurus* skull had been found in Oklahoma, although I did not know it was from the same site that produced the bone fragments Cifelli had shown me. No one had any details about the specimen, although some said the skull was in the hands of a nameless "commercial collector."

About the time that Rich was visiting the site, I was in Tempe, Arizona, where I had been invited to participate in Dinofest International. There, I heard that the North Carolina Museum of Natural History was attempting to buy the skull—and what turned out to be a large part of the skeleton—but the price was hush-hush. One rumor was that the asking price was $400,000. (It eventually sold for much more because of the cost of the work done to clean and cast the bones.) Dr. Dale Russell, who was affiliated with that museum, was quite closemouthed about the price when I asked him. Dale had retired to Raleigh, North Carolina, after serving as the dinosaur paleontologist at the National Museums of Canada (the Canadian version of the Smithsonian Institution). By the time of Dinofest I had known Dale for almost twenty years, and we had published a scientific paper (Carpenter et al. 1997) on a carnivorous dinosaur called *Dryptosaurus* (drip-toe-soar-us), so I was rather surprised at Dale's tight lips about the specimen. I found out later that negotiations were still ongoing.

The specimen fell off my radar until May 1990, when I went to the Black Hills Institute in Hill City, South Dakota, to look at some dinosaur specimens. By then, commercial fossil dealer Allen Graffham owned the Fran fossil, and he had hired the institute to prepare it by removing the encasing rock and repairing any damaged bones. Peter Larson, the institute's president, showed me the specimen and asked if I would be interested in writing a description of it for a scientific journal if the owner approved. Graffham eventually agreed to Philip Currie and me writing a scientific description of the specimen. Phil is renowned for his work on carnivorous dinosaurs, so our collaboration was fitting.

Phil arranged for a family vacation that included a trip to the Black Hills, where we spent several days poring over the material. However, he was busy wrapping up research from a four-year Chinese-Canadian dinosaur project, so the description languished for several years. Neither of us wanted to see this project die, so Phil eventually came down from Canada to Denver, where we hammered out a draft of the manuscript in my tiny apartment. One point on which we disagreed was the evolutionary relationship of *Acrocanthosaurus* to other carnivorous dinosaurs. I argued that the overall shape of the skull, but especially the very peculiar shelf above the eye socket, suggested that *Acrocanthosaurus* was related to *Carcharodontosaurus* (car-care-oh-don-toe-soar-us) from Morocco, based on a recently described skull. Currie, however, argued that the skull and the skeleton showed greater similarities to *Allosaurus* (al-oh-soar-us), an older dinosaur from the Jurassic Period of the American West, and that was how it was placed when the manuscript was eventually published. (See chapter 2 for an update on the relatives of Acro.)

By that time the North Carolina Museum of Natural History owned Fran's skeleton, and Dale Russell gave us permission to publish our description. Although we would have liked to publish it in the Society of Vertebrate Paleontology's international *Journal of Vertebrate Paleontology*, we were unable to do so because this scientifically important specimen had originally been privately owned; many professional paleontologists in North America do not think private individuals should own scientifically important specimens because access is difficult or impossible. Second, the Fran specimen was associated with the Black Hills Institute, which had recently been rocked by controversy surrounding the *Tyrannosaurus* specimen nicknamed Sue. Much has been written about that affair (e.g., Fiffer 2001; Larson and Donnan 2002), most of which is not relevant to my story on Acro. In any case, to avoid objections from colleagues, we looked into submitting the manuscript elsewhere. Phil investigated several scientific journals before deciding on *Geodiversitas*, published in English by the Muséum national d'Histoire naturelle in Paris, France. The article was published at the end of June 2000 (Currie and Carpenter 2000), a decade after I first saw the specimen. Its skull remains the only complete one of this dinosaur to date.

This book originally started as PowerPoint presentations at the annual Acro-Fest at the Museum of the Red River in Idabel, Oklahoma, in the heart of *Acrocanthosaurus* country. The idea for the book was from the director of the museum, Henry Moy. I thank him for his hospitality and the many conversations we enjoyed over dinner during my stays in Idabel. Ms. Jeanette Bohanan, head of programming and outreach, enthusiastically supported Henry's idea, so between the two of them, it was hard to say no. Jeanette acted as liaison and provided me with images and other information needed for this book. I thank Ms. Paulette LaGasse, formerly of the Museum of the Red River, who extended the first invitation to speak at the museum. She unknowingly laid the groundwork for this book. The success of Acro-Fest also depended on the capable hands of Christina Eastep, curatorial assistant at the museum. Thanks also to Quintus Herron for his hospitality during my stay and for having the vision to build a museum in Idabel.

Rich Cifelli (Sam Noble Oklahoma Museum of Natural History, abbreviated as SNOMNH) shared his recollections of the discovery of Fran. He and Collections Manager Jennifer Larsen graciously provided me with access to the dinosaurs from the Antlers Formation. Rich also provided me with the photo of the first Acro discovery in 1940. Timothy Rowe (High-Resolution X-ray Computed Tomography Facility, University of Texas, Austin) gave permission to use the CT scans of the original *Acrocanthosaurus* braincase. Special thanks to the late Allen Graffham, one-time owner of Fran, and Peter Larson (Black Hills Institute of Geological Research—BHI) for the invitation to describe Fran with Philip Currie. Several individuals kindly shared images of *Acrocanthosaurus*, including Peter Larson and Timothy Larson (BHI), Rich Cifelli (SNOMNH), Cephis Hall (Idabel, Oklahoma), James Farlow (Indiana University), and Jerry Harris (Dixie State University).

I owe special thanks to three individuals who took time from their busy schedules to review an early version of the manuscript: Rich Cifelli (who went above and beyond the call of duty by reviewing the manuscript while recovering from surgery), Jerry Harris, and Yvonne Wilson. The manuscript greatly benefited from their input, and this book is partially theirs. Thanks also to Kent Calder, Oklahoma University Press, who championed for this book, and Darcy Wilson, who polished the text through her copyediting skills. Finally, thanks to all those attending the Acro-Fests at the Museum of the Red River for asking such penetrating questions; I hope I've answered them here.

In writing this book I have assumed that you, the reader, have some idea about what a dinosaur is. There have been so many television shows and books on the topic that I assume you have seen at least one. I therefore dive right into the topic of one fantastic dinosaur, *Acrocanthosaurus*. Along the way, I'll share with you where dinosaur research has taken paleontology. Some of this research is based on *Acrocanthosaurus* and some of it on related dinosaurs. Regardless, the underlying goal of this book is to show you how dinosaur paleontology is done and how paleontologists come to their conclusions about those extinct animals.

■

A note about pronunciation: The first time a scientific animal name appears, I convert it to familiar words and syllables rather than dictionary symbols. Technically, this is incorrect, but it is practical because few people know what those dictionary symbols mean. Thus, *Acrocanthosaurus* is "ack-row-can-though-soar-us" rather than "a-krō-kan-tho-'sȯr-əs" and *Tyrannosaurus* is "tie-ran-no-soar-us" rather than "tə-ra-nə-'sȯr-əs."

Acrocanthosaurus
Inside and Out

CHAPTER 1

Discovering Acro

"Acro" is the nickname for the large carnivorous dinosaur *Acrocanthosaurus atokensis* (ack-row-can-though-soar-us ah-toke-en-sis). A cast, or replica, of its skeleton is on display at the Museum of the Red River in Idabel, Oklahoma, and is based on a partial skeleton now at the North Carolina Museum of Natural History in Raleigh (fig. 1.1).

Acrocanthosaurus atokensis was officially named in a 1950 article by J. Willis Stovall and Wann Langston, Jr. The description of the new species was based on parts of two skeletons from Atoka County, Oklahoma; these were the first of five specimens of *Acrocanthosaurus* that we know of today (figs. 1.2 and 1.3). The bones of the first skeleton were reported in late March or early April 1940 by Joe Southern, of the Works Progress Administration, to Stovall, who was then director of the University of Oklahoma Museum. (Later this museum became the Oklahoma Museum of Natural History; now it's the Sam Noble Oklahoma Museum of Natural History, or SNOMNH.) The bones were eroding out of sandstone in what is now called the Antlers Formation. The site looked promising to Stovall, so he returned a short time later with Langston and a small crew from the Works Progress Administration (WPA). The WPA was a federal program created by President Franklin Roosevelt in 1935 to give jobs to the unemployed during the Great Depression. SNOMNH was one of many museums nationwide that benefited from the labor (fig. 1.4).

The excavation site was on the Arnold farm in the southeastern corner of Atoka County, about 7 miles (11.25 km) south-southeast of Farris, Oklahoma. While some WPA workers excavated bones (fig. 1.5), others prospected for

1.1 Cast skeletons of ***Acrocanthosaurus atokensis*** on display A at the Museum of the Red River in Idabel, Oklahoma, and B at the North Carolina Museum of Natural History in Raleigh, North Carolina. *Panel B courtesy of the Black Hills Institute.*

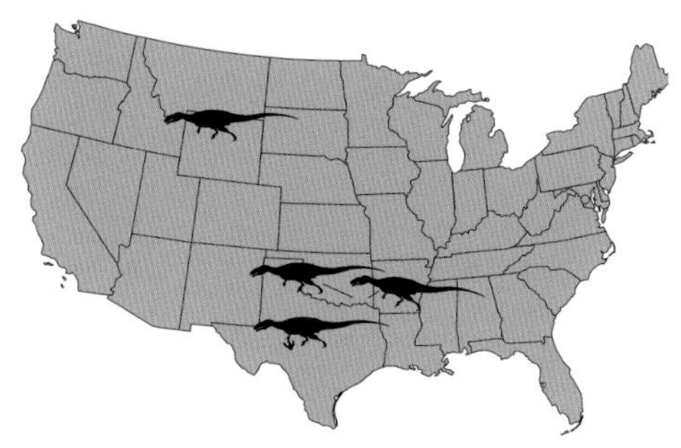

1.2 Map of the known *Acrocanthosaurus* skeletons and main footprint site at Dinosaur Valley State Park. The original two Acros from Atoka County, Oklahoma, are represented by a single silhouette because they are from the same general area. The others are the McCurtain County, Oklahoma, Acro; the Parker County, Texas, Acro; and the Big Horn County, Wyoming, Acro.

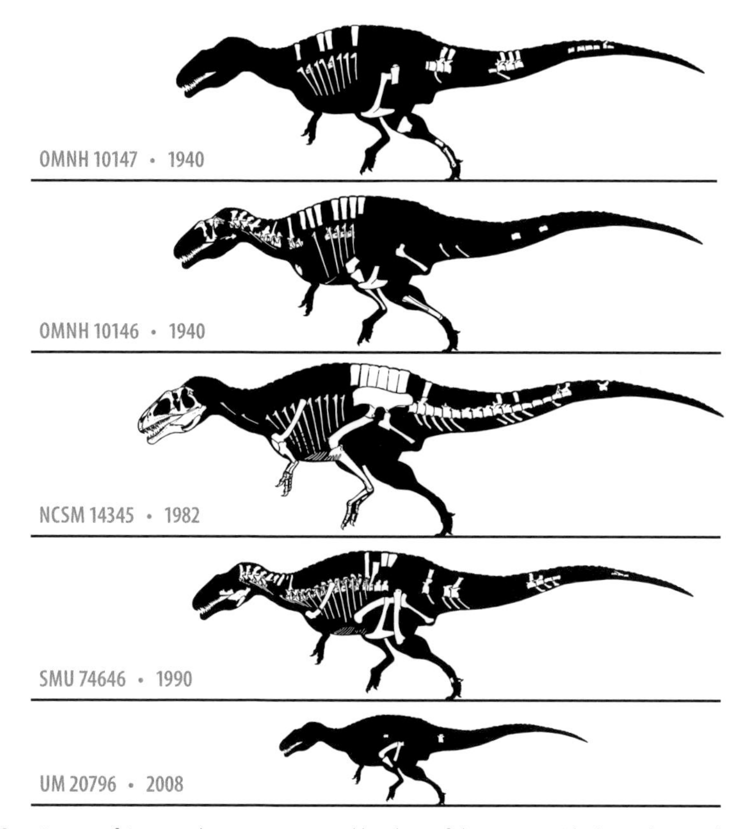

1.3 Specimens of *Acrocanthosaurus* arranged by date of discovery, with the earliest at the top. Bones recovered are shown in white. The museum catalog numbers refer to Sam Noble Oklahoma Museum of Natural History (OMNH), Southern Methodist University (SMU), North Carolina Science Museum (NCSM), and University of Michigan (UM).

A

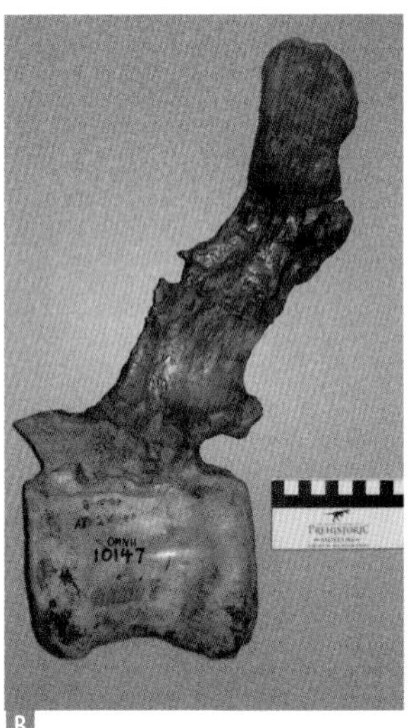

B

C

1.4 A Wann Langston (in tie) at the first *Acrocanthosaurus* site at the Arnold farm. The site was rather muddy because it lay in a stream bank. The man on the left holds a fragment of bone pulled from the mud. Next to him, in a black hat, is Arthur ("Chief") Hutchins, crew chief for the Atoka County WPA crew. B One of the bones Langston studied from this site included a tail vertebra with a tall spine. C Other fossils found in the area included a skull of an herbivore now known as *Tenontosaurus*, although at the time it was unnamed. Scales = 10 cm long (about 4 inches).
Panel A courtesy of Rich Cifelli.

1.5 WPA workers with dinosaur bones wrapped in cocoons of plaster of Paris and burlap (foreground and background), Atoka County, Oklahoma. This method of collecting fragile and broken fossil bones was developed in the early 1880s and is still used by paleontologists today.

other sites in the area by walking the landscape and looking for fossil bone eroding out of the bedrock. Because random digging rarely produces a fossil, prospecting is a more time- and cost-effective means of discovering new sites and specimens. Unfortunately, erosion, which uncovers a fossil, also destroys it. As flowing water erodes the soft rock, the harder, more resistant bone is exposed. However, fossilized bone is usually brittle and full of cracks, so the exposed part tends to break into pieces. These pieces may then be transported downslope by gravity and runoff. By tracing the trail of pieces upslope, a prospector can pinpoint the source of the fossil bone eroding out of the rock. Where the landscape is covered by vegetation, such as in southeastern Oklahoma, erosion is limited mostly to the banks of rivers and streams. It was here that further exploration produced a second skeleton, 0.75 miles (1.2 km) east of the first discovery, on the adjacent Cochran farm.

One specimen was found in relatively soft, yellow sandstone of the Antlers Formation, although many of the bones were encased in a rind of much harder stone. Preservation in a rind of harder stone is common for dinosaur

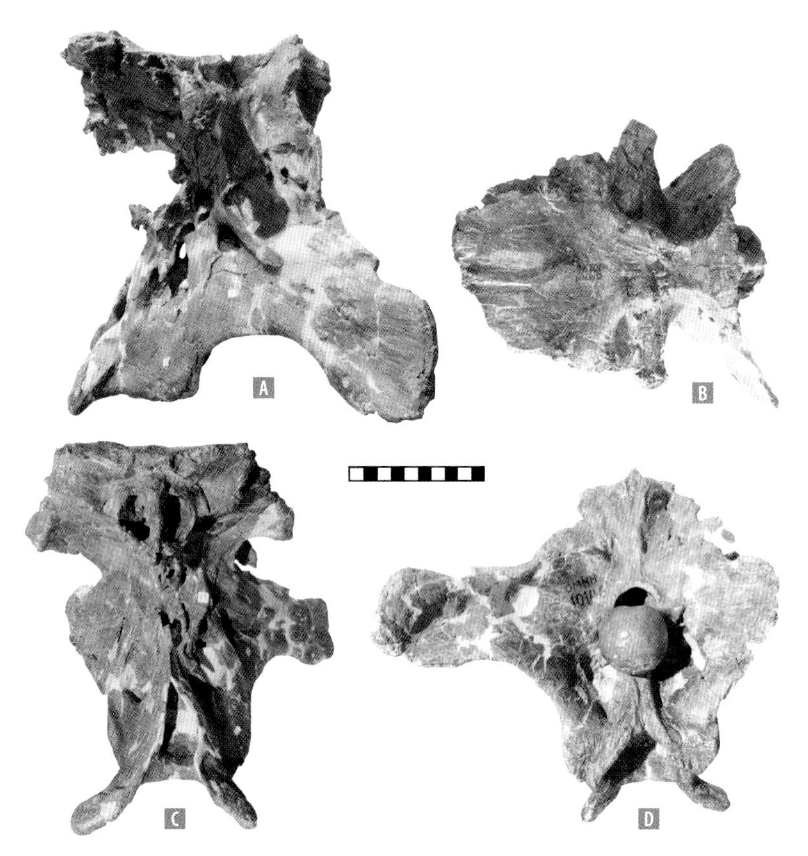

1.6 The all-important braincase of *Acrocanthosaurus* described by Langston shown in A left side view, B top view, C front view, and D rear view. The small white squares in some of the images are Langston's original labels identifying the parts of the braincase. The various openings for nerves and blood vessels can be seen in panels A and C (the two "eyes" are openings for the nerves for smell); the round ball for attaching the skull to the neck vertebrae is seen in panel D, just below the opening for the spinal cord. Scale = 10 cm long (about 4 inches)

bones throughout the world and is due to conditions at the original burial setting (more about this in chapter 3). The other skeleton was slightly more complete than the first and, most importantly, included parts of the skull (fig. 1.6). The tall spines on the vertebrae and the shapes of the pelvic bones, although a little smaller, matched those of the other skeleton, which is how Stovall and Langston were able to show that the two skeletons were of the same type of dinosaur. This specimen was found in dark, reddish-brown, sandy shale, also in the Antlers Formation. The bones lay only a few inches below the surface, and some were heavily damaged by plant roots. Plants often grow into and around fossil bones because of the high concentration of minerals they find there. In Utah I once collected a fossil bone that was little more than a long, brown stain and root mesh. It was only the presence of more recognizable bones inches away, as well as the shape of the root mesh and stain, that enabled me to identify the bone as a dinosaur rib. Fortunately, the Oklahoma bones were pristine, though damaged by erosion.

The success of these early discoveries led the Oklahoma crew to continue prospecting in the area. The result was the discovery of nine other sites containing dinosaur bones, including the skull of a plant-eating dinosaur called *Tenontosaurus* (ten-on-toe-soar-us; see chapter 8). World War II and the subsequent dissolution of the WPA programs brought an abrupt end to further work in Atoka County just as new discoveries were being made. After the war, Langston returned to the University of Oklahoma and described the two specimens that would become *Acrocanthosaurus* in his master's thesis (1947). This thesis was published almost verbatim in 1950, with Langston's thesis advisor, Willis Stovall, listed as lead author. This publication officially recognized *Acrocanthosaurus* as a new type of dinosaur.

It would be more than thirty years before another specimen of *Acrocanthosaurus* was discovered. This third specimen was found about 70 miles (112.6 km) east-southeast of the Cochran site, or about 15 miles (24 km) northeast of Idabel in McCurtain County, Oklahoma (fig. 1.7). Kristi Silvey, a resident of McCurtain County, collected a few bones around 1982. In April 1987 she took Dr. Rich Cifelli, the paleontologist at what was then called the Oklahoma Museum of Natural History, to the site. It was clear to Cifelli that someone had been actively digging up the bones, but he did not know at the time that the site was being excavated by amateur paleontologists Cephis Hall and Sid Love. They'd begun their work around 1983 (fig. 1.8). It was here that Fran was unearthed. The skeleton was found partially in soft, coarse-grained, salt-and-pepper-colored sandstone in the middle of the Antlers Formation area (Miser 1954). The sandstone also contained a lot of plant debris and balls of mudstone. Many of the bones were encased in a hard rind of rock, just like the first Acro specimen excavated by the WPA. Most significantly, the skull was recovered. Most of the bones of this specimen were cleaned of their encasing rock at the Black Hills Institute in Hill City, South Dakota (fig. 1.9), and it was there that I studied the skeleton with my colleague, Phil Currie (Currie and Carpenter 2000). Fran was ultimately purchased by the North Carolina Museum of Natural History in Raleigh, where a plastic cast

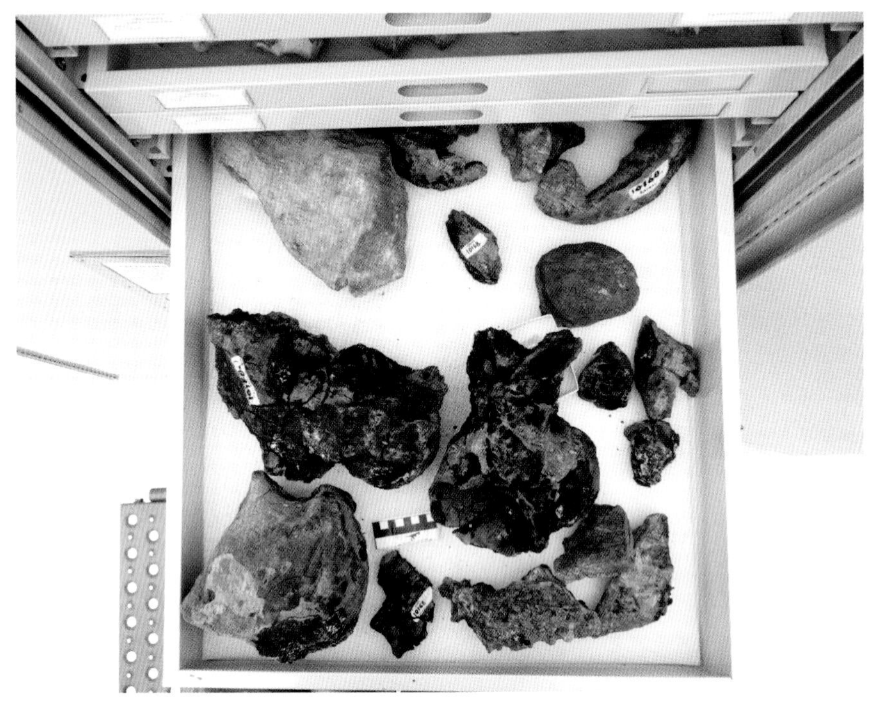

1.7 Some of the first bones of the *Acrocanthosaurus* that would later be named Fran. These bones were collected in the 1980s and are currently in the collections of the Sam Noble Oklahoma Museum of Natural History (as OMNH 10168). The bones visible in this photo are parts of several vertebrae from the back and several unidentifiable fragments

(an exact replica of the skeleton) is on display. Another cast of this same specimen, is displayed at the Museum of the Red River in Idabel.

In 1990 another *Acrocanthosaurus* specimen was collected, this time on the Hobson Ranch in Parker County, west of Fort Worth, Texas. The bones were excavated by Southern Methodist University, Dallas. The specimen was found in medium- to fine-grained sandstone with the sand grains cemented together into concrete by the mineral calcite. This sandstone occurs in the Twin Mountains Formation, which was deposited at the same time (around 110 million years ago) as the Antlers Formation in Oklahoma. The partial skeleton, including bits of the skull, was widely scattered, although some vertebrae lay in sequence as they did in life (fig. 1.10). Cleaning the skeleton was difficult, but when it had been completed, Southern Methodist University graduate student Jerry Harris described it in his master's thesis (1997). Jerry published

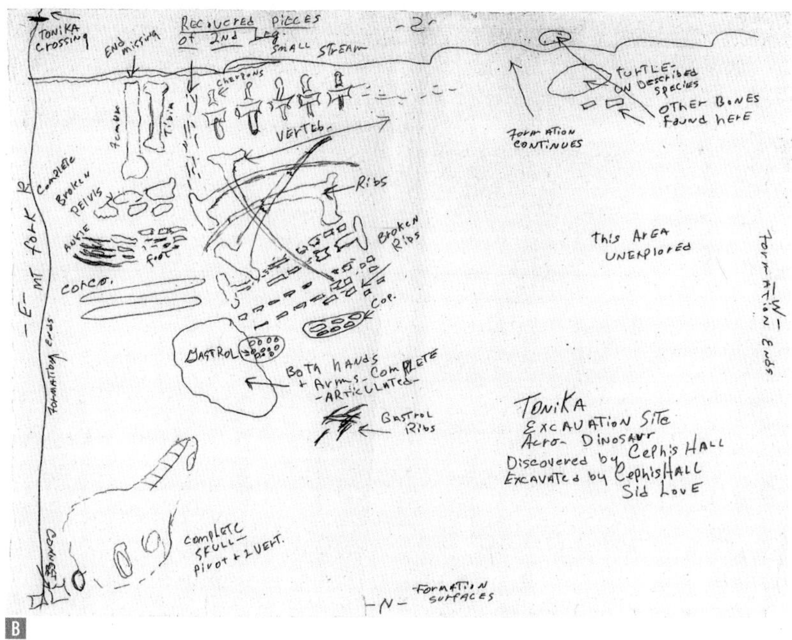

1.8 A Using a backhoe to remove the sediments overlying the *Acrocanthosaurus* skeleton that was later named Fran. B Cephis Hall's map of the distribution of bones discovered during the excavation. The skull is in the lower left corner, and the rest of the skeleton was laying mostly on its right side. *Photo and map courtesy of Cephis Hall.*

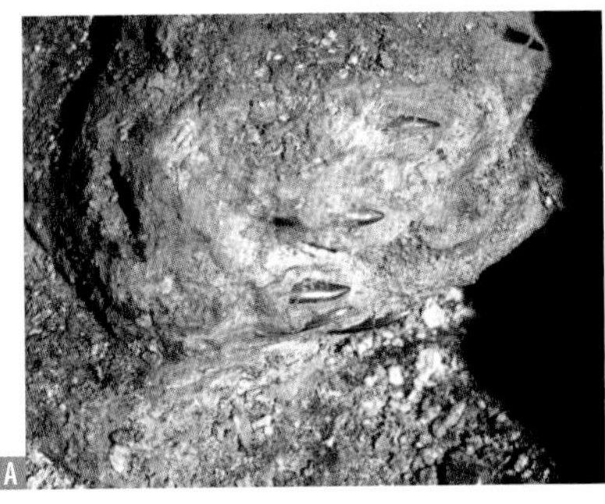

1.9 A Skull of Fran when it was first uncovered by Sid Love and Cephis Hall. The upper teeth of the snout are visible. B The skull when it was received by the Black Hills Institute. The rear part of the skull was separate. C Removal of the encasing rock (preparation) by fossil technician (preparator) Terry Wentz involved a variety of tools, including pneumatic scribes, needles, X-Acto™ knives, and air abrasion, which blows an abrasive at high pressure. D Once preparation was completed, the individual bones were separated so that each bone could be molded and cast and the entire skull reconstructed in its uncrushed state. *Panel A courtesy of the Museum of the Red River; panels B–D courtesy of the Black Hills Institute.*

his thesis a year later, two years before Currie and I published our description of Fran. The specimen Jerry studied was important because it contained parts of the skeleton that had not been available to Stovall and Langston, including parts of the skull and lower jaw, neck vertebrae, and shoulder blade.

The fifth and most recently collected *Acrocanthosaurus* specimen consists of a handful of bones from the Cloverly Formation of northeastern Big Horn County, Wyoming. The specimen was found in 2008 by paleontology student Mike D'Emic from the University of Michigan. This partial skeleton is from an *Acrocanthosaurus* that was about three-quarters grown (D'Emic, Melstrom, and Eddy 2012) and thus represents the smallest of the five known specimens.

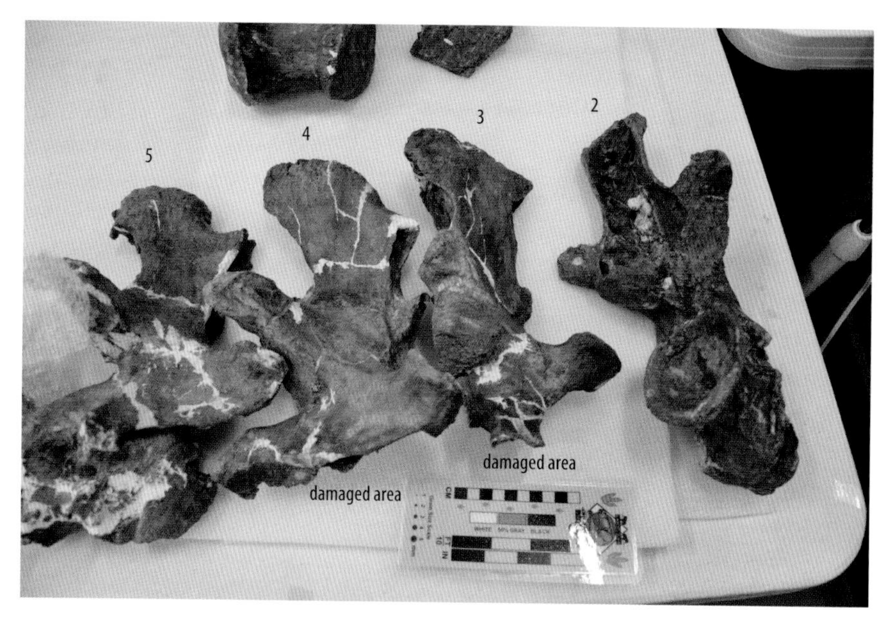

1.10 Right side of neck vertebrae 2 through 5 of the specimen from the Hobson Ranch in Texas. Note the hatchet-shaped spines that project upward. Their large size provides a strong anchor for the neck muscles. Scale in centimeters. *Photo courtesy of Jerry Harris.*

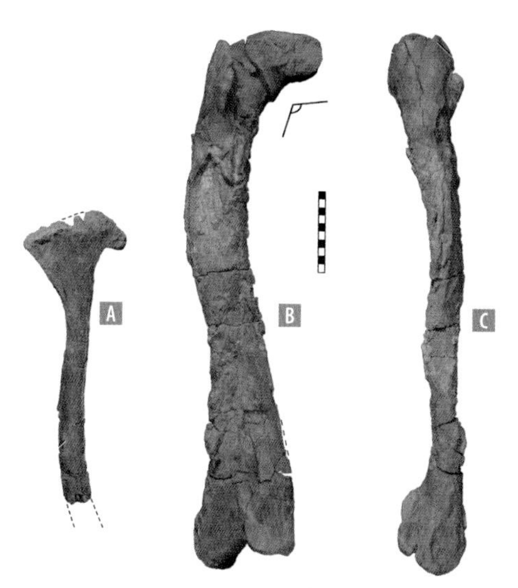

1.11 Two of the bones of the Wyoming *Acrocanthosaurus.* A Part of the pelvis (pubis). B Right thigh (femur) in front and C side views. The angle of the femur head to the shaft is similar to that of the Hobson Ranch specimen and is one of the main reasons this specimen is considered to be an *Acrocanthosaurus.* Note the severe crushing seen as flattening of the femur in side view. *Colorized images reprinted from Michael D. D'Emic, Keegan M. Melstrom, Drew R. Eddy (2012), with permission from Elsevier.*

The find occurred in dark, brick-red, silt-rich mud rock (mudstone) with green and yellow splotches caused by subtle chemical differences in the rock. The fossil bones were badly crushed from compaction of the original layers of mud during the process of becoming rock (fig. 1.11). This *Acrocanthosaurus*, about 1,000 miles (1,600 km) northwest of the Oklahoma specimens, is the northernmost Acro find. Fossil footprints that may have been made by *Acrocanthosaurus* suggest that the species might have lived as far north as southern Canada (Currie, personal communication, 2015). Big teeth and other isolated bones in Lower Cretaceous rocks (145–100.5 million years old; fig. 1.9) from a scattering of localities, including Utah and Maryland (Lipka 1998), have been tentatively identified as *Acrocanthosaurus*, but a re-examination of the specimens by one of Wilson's former students, Mike D'Emic, casts doubt on these identifications (D'Emic, Melstrom, and Eddy 2012).

Five partial skeletons may not seem like much, but for dinosaur paleontologists they are a treasure trove. Most of the dinosaurs we study are known from far fewer bones. As I will show you in the rest of the chapters, these five specimens have revealed a lot about *Acrocanthosaurus*, and I hope that along the way you'll be able to imagine Acro as a living, breathing creature, not just a collection of old bones.

CHAPTER 2

Who Is *Acrocanthosaurus?*

When Langston studied the first *Acrocanthosaurus* specimens in the late 1940s, he knew the bones were those of a carnivorous dinosaur. All carnivorous dinosaurs belong to a group called theropods, though a few members of this group have at various times appeared to have given up chasing and killing prey to become vegans, eating various plants.

Two of the most famous carnivorous theropods are *Tyrannosaurus* and *Allosaurus*, which are well known scientifically from numerous fairly complete skeletons. We know that they were carnivores based on their long, narrow, blade-like teeth, which are well adapted for slicing flesh. Neither of the two specimens of Acro available to Langston at the time had teeth, but he was still able to determine that the skeletons were those of a theropod because the skull bones more closely matched those of *Tyrannosaurus* and *Allosaurus* than those of any other dinosaur. This was especially true for the bones surrounding the brains in these three dinosaurs. Other similar features were the *Allosaurus*-like ball-and-socket neck vertebrae (in *Tyrannosaurus* and many other dinosaurs these vertebrae are flat-faced); the spines on the vertebrae, which are similar to those of *Allosaurus* in shape and structure, though much taller (fig. 2.1); and the pelvic bones, which are similar to those of *Tyrannosaurus* and especially *Allosaurus*.

Despite these similarities in Acro to *Tyrannosaurus* and *Allosaurus*, there were also major differences, which convinced Langston (1947) and Stovall and Langston (1950) that they were dealing with a new dinosaur. When coining new names, scientists are guided by the International Code of Zoological

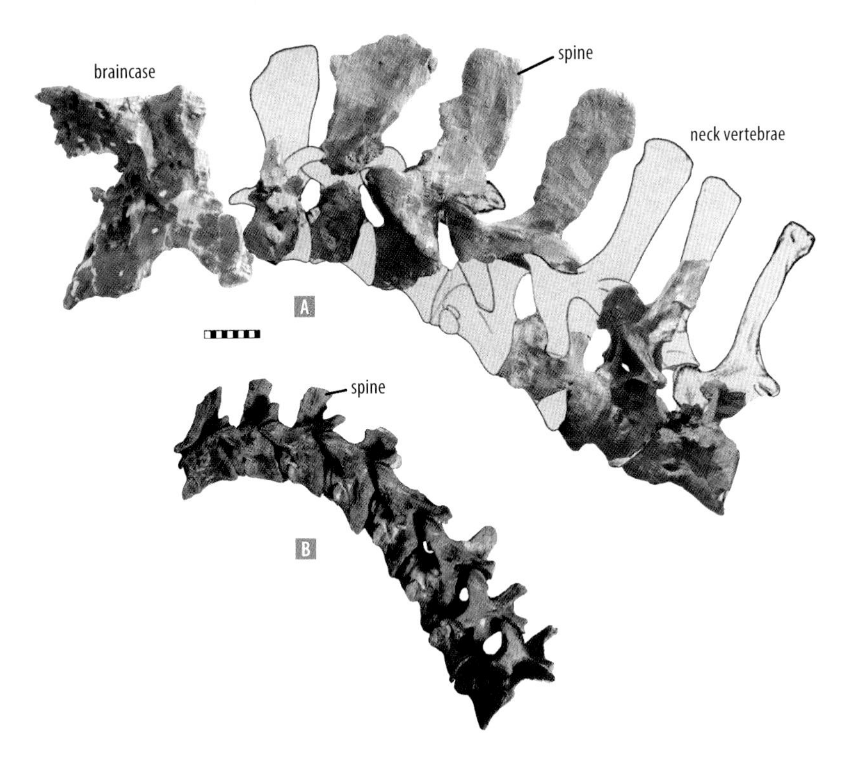

2.1 **A** The braincase and neck vertebrae that Wann Langston used in his master's thesis description of *Acrocanthosaurus*. Note the uniquely tall spines, called neural spines. **B** The neck vertebrae of *Allosaurus*, which Langston thought was related to *Acrocanthosaurus*. The proportionally shorter neural spines in *Allosaurus* represent the more common condition among theropods. *Allosaurus vertebrae courtesy of Gregory Paul.*

Nomenclature (ICZN), which ensures that a new scientific name is valid. The code requires that the name have two parts: the first is the genus name and the second is the species name. Stovall and Langston named this new dinosaur *Acrocanthosaurus atokensis*. In creating the genus name, they combined the Greek words *akra*, meaning "high," *akantha*, meaning "spine," and *sauros*, meaning "lizard," to construct the name "high-spined lizard" in reference to the particularly tall spines on the vertebrae (fig. 2.2). The species name refers to Atoka County, where the specimens were found (the root *ensis* means "from").

Another ICZN requirement is that one specimen, called a holotype, be selected as the name-bearer. This one specimen becomes the standard for the species; all similar specimens must be compared with it to determine whether they belong to the same species or a different one. The holotype

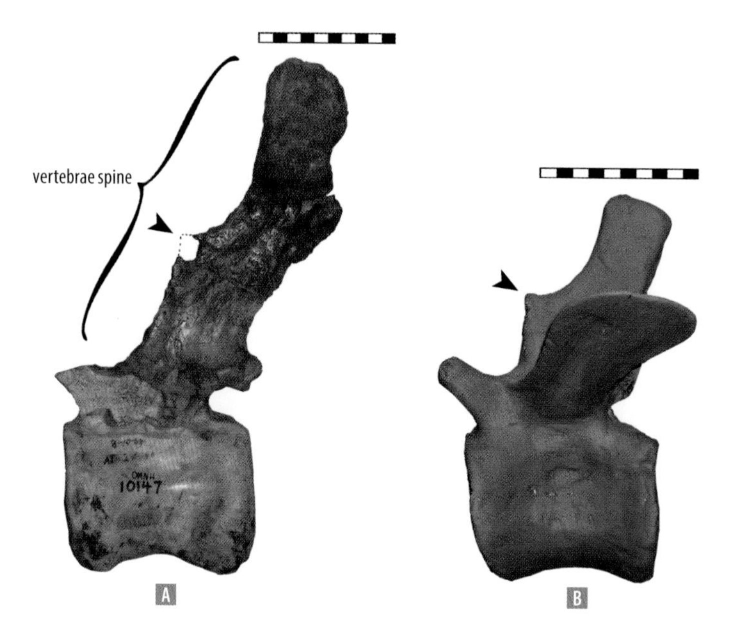

2.2 Acro's tall neural spines can be appreciated by comparing **A** the eleventh tail vertebra of *Acrocanthosaurus* with **B** the same vertebra of *Allosaurus*. Because the *Acrocanthosaurus* vertebra is much bigger, the two vertebrae have been scaled to the same length of the vertebral body to show the proportionally different spine height. The arrows point to the same spur on both spines (damaged in *Acrocanthosaurus*). This landmark shows that the greater height of the spine in the *Acrocanthosaurus* is not solely due to the addition of bone at the top of the spine. The spur and the step it produces above it are characteristics Stovall and Langston used to suggest that *Acrocanthosaurus* was related to *Allosaurus*. Scales are 10 cm long (about 4 inches).

must have at least one unique characteristic that defines or diagnoses the new species and makes it different from all others. If more unique characteristics can be identified, they will strengthen the case for the specimen belonging to a distinct species. Stovall and Langston had two specimens from which to work. They selected the second specimen, which had skull parts, as the holotype and made the other a paratype. A paratype is a supporting specimen that complements the holotype in rounding out the description of the new species. Usually it contains parts missing in the holotype. In the case of these first two *Acrocanthosaurus* specimens, the foot bones of the paratype were used in the scientific description because they were missing in the holotype.

Stovall and Langston (1950) gave a rather long list of characteristics in the diagnosis of *Acrocanthosaurus*, not all of which are unique to the species. For example, they mentioned that it was a "carnivorous saurischian of gigantic size and heavy proportions" (p. 700). Even in 1950, these terms were true of other dinosaurs, such as *Tyrannosaurus* (named by Osborn 1905). Also, some carnivorous dinosaurs that have been discovered since the original description have characteristics that were originally thought to be unique to Acro. One of these characteristics was the prominent ball-and-socket connection of the first few neck vertebrae. Such characteristics no longer form part of the diagnosis for *Acrocanthosaurus,* but they are still helpful in determining how closely related *Acrocanthosaurus* is to other theropods.

I now give you a scientific definition of *Acrocanthosaurus atokensis*. Instead of "Large theropod with elongate neural spines that are more than 2.5 times corresponding presacral, sacral and proximal caudal lengths of the centra" (p. 211), as Phil Currie and I gave it in our scientific description (Currie and Carpenter 2000), I'll give it in plain language. However, I'll have to use the scientific names of bones because most bones have no common name.

I'll discuss and illustrate the bones of *Acrocanthosaurus* in chapter 4; for now, just know that its skull and lower jaw are made up of multiple bones. One of these, on the back of the skull, is a wedge-shaped bone called the supraoccipital; it sits above the opening through which Acro's spinal cord passed to the brain. The supraoccipital has two V-shaped ridges that end in little flat surfaces, and the neck muscles that tilt the head back were attached to these. Also along the back of the skull is a vertical bone called the quadrate, to which the lower jaw attaches in a hinge-like joint. On the inner side of the quadrate is a pocket that was probably connected in some manner to the air system that was spread throughout the body. In the lower jaw, one of the bones that makes up the rear half, called the surangular, has a ridge or shelf to which jaw-closing muscles attached. (Chapter 6 discusses more about this and the air system.) This ridge or shelf has a localized thickening, or knob ("bump"). Below the shelf is a large opening through which a nerve passed.

All of the vertebrae of the neck, back, and pelvis have depressions and smaller openings into which some of the air system passed. One of the most distinctive features, and what gave *Acrocanthosaurus* its name, are the tall spines that project upward (I discuss these at greater length in chapter 5). The height of these spines is about two and a half times the length of their individual vertebrae. Those on the neck also have an odd, triangular spur that inserts into a pocket on the backside of the next vertebrae in front. Finally, the vertebrae from the middle of the tail have a secondary flange that projects out to the side.

Note that it took me more than 250 words here to define Acro, whereas it took Currie and me only 80 words using technical language. That is why scientists use technical words that can sound like gibberish.

Because the bones of *Acrocanthosaurus* are more similar to those of *Allosaurus* (figs. 2.1 and 2.2) than of any other theropod that Stovall and Langston knew about, they placed it in the same biological family as *Allosaurus*, Antrodemidae (an-trow-deem-eh-day), now called Allosauridae (al-oh-soar-eh-day). In doing so, they were implying that the two dinosaurs were more closely related to each other than to any other theropod, such as *Tyrannosaurus*. But all this was to change in the 1990s with the discovery of a large theropod skull in Africa, of all places. As is common in paleontology, a new discovery can change what we think we know. In this case, the discovery of the skull in southern Morocco showed that *Acrocanthosaurus* was actually most closely related to a dinosaur called *Carcharodontosaurus* (car-care-oh-don-toe-soar-us), which was not known well in the 1950s.

This new theropod skull was found in the Kem Kem beds in the Kem Kem Basin along the border between Morocco and Algeria. In the 1990s this area was known to be rich in dinosaur fossils. Around the world, thousands of dinosaur teeth and bones from the Kem Kem were turning up for sale at rock and mineral shows. Dinosaur paleontologist Paul Sereno from the University of Chicago figured that there must be more from the area than just isolated bones and teeth, and the discovery of a 5-foot 3-inch (160-cm) skull of *Carcharodontosaurus* (Sereno et al. 1996) exceeded his wildest dreams. *Carcharodontosaurus* was a previously poorly known theropod named by German paleontologist Ernst Stromer (1931) from a few bones and teeth he collected in the Western Desert of Egypt. In comparing the new skull of *Carcharodontosaurus* with those of other theropods, Sereno and his colleagues also noted that *Carcharodontosaurus* was related to another large theropod, *Giganotosaurus* (jig-ah-no-toe-soar-us) from Argentina.

This discovery had important implications for *Acrocanthosaurus*. Contrary to what Stovall and Langston (and other paleontologists) thought, the closest relatives of *Acrocanthosaurus* did not include a dinosaur from North America (*Allosaurus*) but, rather, were dinosaurs from Africa and South America (fig. 2.3). This startling conclusion has since been verified by several other scientific studies, such as the one by Steve Brusatte and Paul Sereno (2008). *Acrocanthosaurus* is now placed with *Carcharodontosaurus* and *Giganotosaurus* in the family Carcharodontosauridae (car-care-oh-don-toe-soar-eh-day), along with several more recently named theropods. Within this family group, or "clan,"

2.3 A Skull of *Acrocanthosaurus* compared with the skulls available to Stovall and Langston in 1950: B *Allosaurus* C and *Tyrannosaurus*. The *Acrocanthosaurus* skull more closely resembles that of *Allosaurus* than it does the short, deep skull of *Tyrannosaurus*. The discovery of two long, low skulls in the 1990s, D *Carcharodontosaurus* and E *Giganotosaurus*, revealed that they had features in common with *Acrocanthosaurus*, such as the long, tapering snout and flat shelf above the eye socket. Abbreviations: e = eye socket, n = nostril opening.

if you will, *Carcharodontosaurus* and *Giganotosaurus* are more closely related to each other than either is to *Acrocanthosaurus*. The family Carcharodontosauridae, also known as carcharodontosaurids (car-care-oh-don-toe-soar-ids), is distantly related to *Allosaurus*. Together, *Allosaurus* and carcharodontosaurids make a larger group of theropods, the Allosauroidea (al-oh-soar-oy-dee-ah). Confused? Another way of thinking about this is to consider *Acrocanthosaurus* as a half-sister to the sisters *Carcharodontosaurus* and *Gigantosaurus*; thus all three belong to the same family, whereas *Allosaurus* would be more like a second cousin (more distantly related). All four dinosaurs—*Acrocanthosaurus*, *Carcharodontosaurus*, *Giganotosaurus*, and *Allosaurus*—belong to the greater clan, the Allosauroidea.

How the carcharodontosaurids spread to Africa, South America, and North America within a few million years can be explained by the continents still

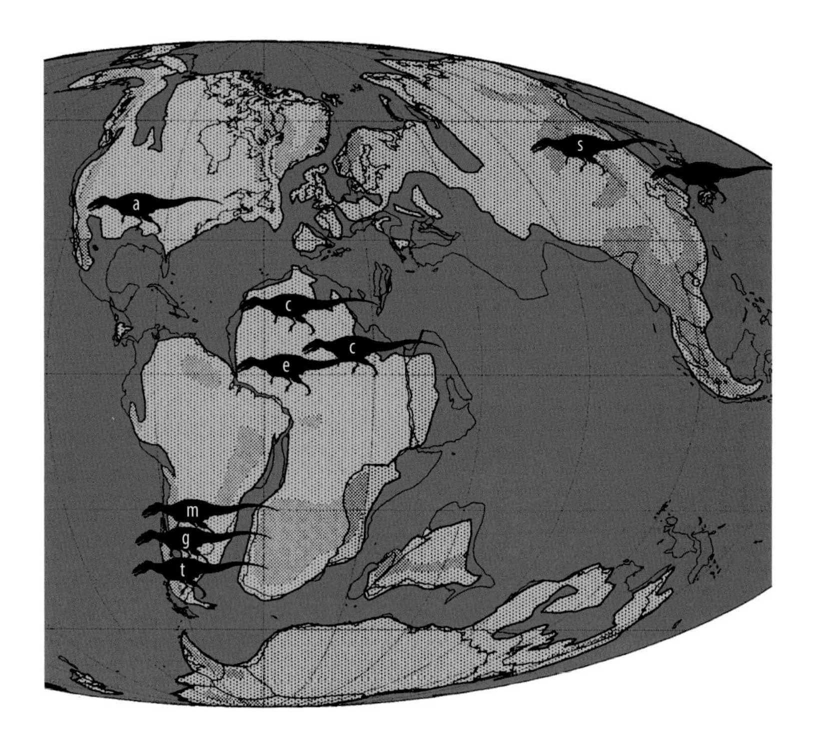

2.4 Distribution of *Acrocanthosaurus* and some of its relatives, the carcharodontosaurids, on a paleogeographic map of the end of the Early Cretaceous showing land (green) and sea (blue; a = *Acrocanthosaurus*, c = *Carcharodontosaurus*, e = *Eocarcharia*, g = *Giganotosaurus*, m = *Mapusaurus*, s = *Shaochilong*, and t = *Tyrannotitan*). Not all of these dinosaurs are discussed in the text, but they are included here to show the distribution and diversity of the relatives of *Acrocanthosaurus*; the silhouette on Japan is an unnamed species known from teeth. Two other dinosaurs, *Neovenator* from England and *Concavenator* from Spain, may or may not belong to the same family as *Acrocanthosaurus*. The most recent study suggests that they are different enough to be in a separate family called the Neovenatoridae. I've chosen the more progressive approach, so these are not shown on the map.

being joined or at least fairly close together. Although we tend to think of the continents as being stable, earthquakes show that is not true. In fact, if we plot the locations of earthquakes, which are the result of rocks suddenly jerking past each other along giant cracks called faults, we can see that earthquakes do not occur randomly. Instead, they outline large areas of the earth's surface into regions called plates. These plates float on a layer of semimolten rock and get moved around by the churning motion of the deep interior of the earth. This movement is about 0.5–1 inch (1.25–2.5 cm) per year. How much have plates moved in your lifetime? A truly wild fact is that airplanes now have to fly 8 feet (2.4 m) farther when crossing the Atlantic than the first transatlantic flight did in 1919. Over geologic time, measured in millions of years, the plates with the lighter continents riding on top can move around a lot. At various times in the geologic past, the continents were closer to each other and at times were even joined into larger continents. At such times carcharodontosaurid dinosaurs and other land animals could move from continent to continent.

So where did *Acrocanthosaurus* come from? A clue may have been uncovered by Oliver Rauhut, a dinosaur paleontologist at the Humboldt Museum of Natural History in Berlin. Rauhut was studying Late Jurassic theropod bones that were collected shortly before World War I by that museum in what was then German East Africa (modern Tanzania). Among the drawers of bones he found three tail vertebrae that close inspection revealed to be unlike anything else in the collection. He was surprised to note that their closest match was to the peculiar middle-tail vertebrae of *Acrocanthosaurus.* (Remember, these particular vertebrae have the unique second flange that was used in the definition.) What was odd was that these vertebrae were 35 million years older than *Acrocanthosaurus* and in the same age range as *Allosaurus.* If anything, he would have expected the African vertebrae to more closely resemble those of *Allosaurus.* Rauhut (2011) concluded that these vertebrae represented the earliest known carcharodontosaurid, which he named *Veterupristisaurus* (vet-err-up-riss-tee-soar-us). Rauhut suggested that *Veterupristisaurus* and *Acrocanthosaurus* were closely related; in fact, he called them "sisters."

If—and this is a big "if"—*Veterupristisaurus* was an ancestor to *Acrocanthosaurus*, then its descendants immigrated to North America either by way of Europe or by way of South and Middle America. Whichever way they went, some swimming between islands was required because at that time (Late Jurassic and Early Cretaceous) most of Europe and Mexico consisted of a series of islands, much like the Philippine archipelago does today (fig. 2.4). We do not know when the ancestor of *Acrocanthosaurus* first set foot in what is now Texas and Oklahoma; we know only that *Acrocanthosaurus* was present there by around 110 million years ago.

CHAPTER 3

How to Fossilize an *Acrocanthosaurus*

We don't know what killed Fran or the two specimens that Langston described. In fact, we rarely know the cause of death for most dinosaur specimens because all that remains are the bones. Injury to the soft tissue—everything in the body that is not bone or teeth—rarely leaves any evidence on bone. Death by disease and death by starvation will look the same when all that remains are scattered bones. It is also nearly impossible to distinguish bones of an animal that was killed by a predator from those of an animal that died of other causes and was later scavenged. In addition, dinosaur skeletons are rarely complete from nose to tail because a carcass is a free lunch to any carnivore within smelling distance. Carnivores are notoriously lazy. Why expend the energy stalking, chasing, and fighting with a struggling prey, not to mention risking bodily injury, when a free meal is just lying there?

Scavengers are hard on skeletons because they tear apart bones at their joints. Fortunately, dinosaur scavengers were not as destructive to individual bones as modern mammalian scavengers are. Dogs, for example, love to gnaw on bones and can fracture a whole bone into small pieces, whereas carnivorous dinosaurs lacked the crushing molars of mammalian scavengers. Instead, they had a mouth full of teeth resembling steak knives. These teeth are better at stabbing and slicing through flesh than they are at crushing bones. Still, the teeth of a scavenger inevitably would strike bone, leaving long gouges as the tooth scraped the bone surface. Teeth can also break off at this time, and for this reason isolated teeth from scavenging carnivorous dinosaurs are commonly mixed among the bones of their meals. The teeth of modern crocodilians can also be found around scavenged carcasses, which shows that scavenging has long been a tradition among these reptiles.

I know that Fran was scavenged because V-shaped gouges made by steak-knife teeth are present on some of the bones. Such damage is different from the U-shaped gouges that are accidentally made by digging tools or by the lab tools used to remove rock from fossil bones. It seems likely that the scavenger was another *Acrocanthosaurus* because loose *Acrocanthosaurus* teeth were found at the site. This dinosaur scavenger was in a race for the carcass with other scavengers and decomposers, especially insects and bacteria. That sickly smell associated with roadkill is caused by putrescine (pute-re-seen) and cadaverine (cad-ave-er-een), two chemicals produced by the bacterial breakdown of organic compounds called amino acids. (Incidentally, putrescine and cadaverine occur in bad breath, so brush your teeth!) These chemicals can be toxic in high concentrations. Maggots and other insect larvae today can neutralize these toxins and presumably did so in the ancient past. By the time a carcass is covered with maggots, vertebrate scavengers either have had their fill or will turn their nose up at the carcass.

Eventually, Fran was reduced to scattered bones with little tissue left. It may have taken weeks or, more likely, months (because of its sheer size) to reach this stage. If the bones had not been buried by this stage, they would have started to crumble as the collagen within them started to decompose. Anna "Kay" Behrensmeyer (1978), a paleobiologist at the Smithsonian Institution, has studied the breakdown of bone exposed to the environment. She noted that the first sign of a bone's destruction is the appearance of cracks on the bone surface. These cracks become larger and more abundant over time, and eventually the bone is reduced to small slivers. Termites are also known to cause havoc to bones by burrowing up from the ground and then throughout bone, although whether they do this for housing or to feed on the organic material is unknown. Other damage can occur if the bones are buried in acidic soil. Acids produced from decaying plant material can slowly dissolve the mineral part of bone, making it easier for bacteria to get at the many nonmineralized parts of the bone. Fortunately, in the case of the Fran, none of these extremes happened. Instead, the bones were buried by sediment carried by flowing water (fig. 3.1). The conditions that enable fossilization are rare, so Acro fossils are scarce because only a small percentage of the Acros that ever lived were fossilized.

3.1 Scattered bones of an *Acrocanthosaurus* lying in a riverbed. Flowing water can scatter bones, but it can also move sand and gravel, which will bury them and increase their chance of becoming fossilized.

Figuring out the environment where Fran was buried involved some detective work on my part. When I first saw Fran at the Black Hills Institute, I noticed that the sedimentary rock still encasing parts of her was a mixture of fine sandstone and dark mudstone (a rock made from mud). Also, pyrite was found on and in the bones. Pyrite, a mineral made of iron and sulfur atoms, was the most important clue because it forms in oxygen-poor environments, such as stagnant water. This suggested to me either a swamp or a stagnant pond, but which was it? Photographs of the site by Cephis Hall showed that the bones originally lay in a lens-shaped deposit of charcoal-gray shale (figs. 3.2 and 3.3). The thin layers made of clay particles indicate deposition in quiet water, and the lens shape indicates an abandoned river channel. From

3.2 Close-up of the dark, organic-rich shales of an ancient oxbow lake. Note the thin layers, which are called lamina. The yellow stains are jarosite and other iron and sulfur minerals that commonly form in organic-rich deposits. Black flakes of coalified wood and charcoal are also present from vegetation that existed nearby.

these lines of evidence I conclude that Fran was buried in an oxbow lake. The active river channel was close and would occasionally spill out of its banks through a gap during large floods to deposit sediments called crevasse splay deposits into the abandoned channel. This deposition of sand happened at least three and possibly four times, based on the sandstone layers. Did Fran die in this pond, or was her body transported there during a flood? Did she die during a drought, or did she drown? Unfortunately, nothing about her skeleton can provide the answer.

Many of Fran's bones were also encased in rinds of hard rock, called concretions (fig. 3.4). These concretions were made of calcium carbonate, which is the same mineral that makes limestone. Ground-up limestone that has been baked is used to make construction concrete, so Fran was literally encased in a natural concrete. As we shall see, calcium carbonate is important to our story about the fossilization of Fran.

First, what is calcium carbonate? A single molecule—a collection of atoms of different elements—of calcium carbonate has the chemical formula $CaCO_3$, which means one atom of calcium (Ca) is attached to one atom of carbon (C), which is in turn attached to three oxygen (O) atoms. The carbonate (CO_3) part of the molecule is actually a carbon dioxide (CO_2) molecule with an extra atom of oxygen. Calcium is very common in the earth's crust, where it is a

3.3 Site of Fran as viewed from across a gully that leads to the Mountain Fork River (out of view to the left) and interpretive sketch. A faint, handwritten note on the right of the photo says "bone bed" with an arrow pointing to the bone-bearing level. The lens shape of the darker sediments shows the position of an ancient, abandoned river channel (tens of millions of years before the existence of the Mountain Fork River) that contained the bones. Periodic floods spilled sand from the active channel somewhere to the left through a breach in the natural levee. This sand eventually covered the former channel (capping sandstone). The thin layer of Ice Age gravels is a much more recent deposition by an older version of the Mountain Fork River. *Photo courtesy of the Museum of the Red River.*

3.4 Many of Fran's bones were found covered with a thin layer of white calcium carbonate concretion. **A** Lower arm, wrist, and finger bones. **B** Shoulder blade and upper and lower arm bones. *Pictures courtesy of the Museum of the Red River.*

component of various minerals, such as gypsum and fluorite. Erosion and weathering breaks down these minerals and liberates the calcium atoms, Ca^{2+}. (The "2+" means the atom has a positive electric charge and wants to stick to atoms with a negative charge because opposite charges attract each other.) The most common molecule that calcium sticks to is carbon dioxide (CO_2), which makes calcium carbonate through a series of chemical steps. Geologists originally thought that calcium carbonate formed spontaneously whenever calcium and carbon dioxide were brought together. As chemists discovered, the problem with this idea, or hypothesis, is that the process by itself is very slow, making it difficult to account for thick calcium carbonate or limestone deposits found in the earth.

A breakthrough came when scientists noticed that calcium carbonate tends to occur in the presence of certain microbes, whether they are in the soil, such as around the roots of plants; in water (CO_2 dissolves naturally in water anywhere that water and air are in contact with each other); or in the human body, where deposits such as kidney stones can form. These observations were replicated in the lab (e.g., Hammes and Verstraete 2002), so the evidence is conclusive that bacteria and certain other microbes can rapidly deposit calcium carbonate as a solid mineral. This process, in fact, is many, many times faster than spontaneous chemical formation. Bacteria form calcium carbonate by using the atoms of calcium to get rid of their bodily waste, mostly carbon dioxide. Think of it as bacteria attaching their "poop" of carbon dioxide to atoms of calcium. When this happens, molecules of calcium carbonate form, clump together, and precipitate out of the water as a solid, thus removing the bacterial waste. So much of this calcium carbonate precipitates out that it can form a solid bed of limestone.

In some soils, calcium carbonate can also cement the surrounding sediment into rounded lumps called concretions. As many fossil collectors know, concretions are famous for the fossils that may occur within. Experimental work has shown that concretions often form around buried, decaying tissues that are rich in bacteria. As the calcium carbonate forms because of the bacteria "poop," it cements the surrounding sediments around the decaying material. Concretions can form irregular lumps, especially on fossil bone. Such masses were found on many of Fran's bones and show where tissue was still decaying after the bones were buried (fig. 3.4).

Reading about how concretions and limestone form made me wonder if bacteria might play a role in the most common type of bone fossilization, which is petrifaction. All the books that I grew up with said that that petrifaction was a molecule-by-molecule replacement of bone by mineral. Even my teachers said that, but no one seemed to know how this happened. The problem with this replacement model is that it did not jibe with the numerous chemical studies of fossilized bone that showed bone mineral, hydroxyapatite, was still present. It seemed to me that bacteria must play a role, but how could I prove it?

I decided to bring together bacteria, dissolved calcium carbonate, and bone to see if an artificially fossilized bone could be produced through the deposition of dissolved mineral within the bone (a process called permineralization; Carpenter 2005). I bought a cow bone sold for dogs in a grocery store and cut it into 1-inch (2.5-cm) cubes. I left these sitting out overnight to start the process of decay by letting bacteria colonize and grow. The next day, I buried these bones in a glass funnel full of play sand. Next I took some limestone powder and boiled it in distilled (mineral-free) water to dissolve the calcium and release the carbon dioxide. When it cooled, this water solution was dripped through the sand at about one drop per second and was allowed to drip out the bottom after percolating through the sand (fig. 3.5). The bone in the sand was an imitation of bone buried in a riverbed, where a steady supply of groundwater percolates through. Because some bacteria don't like light, I wrapped the sand funnel with aluminum foil, which simulated the sand being underground and out of the light. The water dripping out the bottom was smelly, indicating the presence of decay bacteria.

After letting the water drip through the sand for a month, I removed the first sample. I was surprised to find a thin rind of sand grains cemented to the

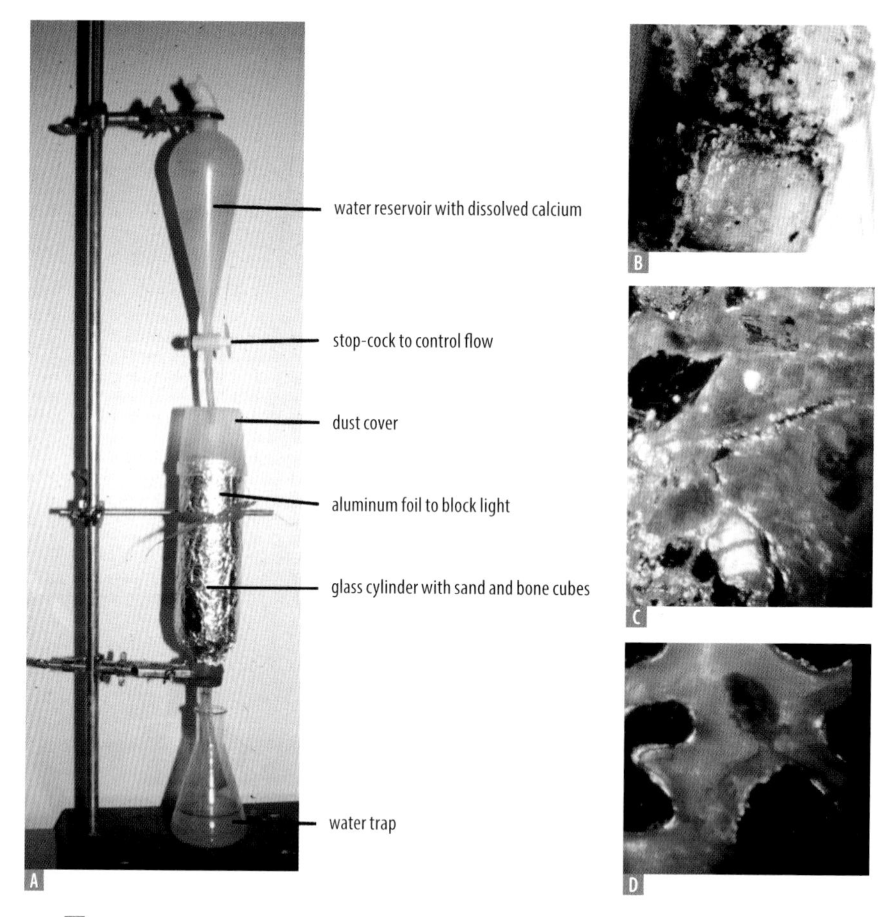

3.5 **A** The experimental apparatus I used to learn whether bacteria play a role in the fossilization of bone. The water reservoir at the top contained a solution of water and dissolved calcium. This solution was allowed to drip through a glass cylinder of sand and cubes of fresh bone. **B** Bacteria used the atoms of the dissolved calcium to get rid of their carbon dioxide waste, forming calcium carbonate as a solid, which "glued" the sand grains together on the bone surface as a rind. This is the same process that formed the calcium carbonate rind on the surface of the bones of Fran. **C** Within the bone, calcium carbonate deposits (arrows) eventually formed once some tissue had decayed and left room for them. **D** In a sterile version of the experiment, bacteria could not grow because of added bleach and no calcium carbonate was deposited.

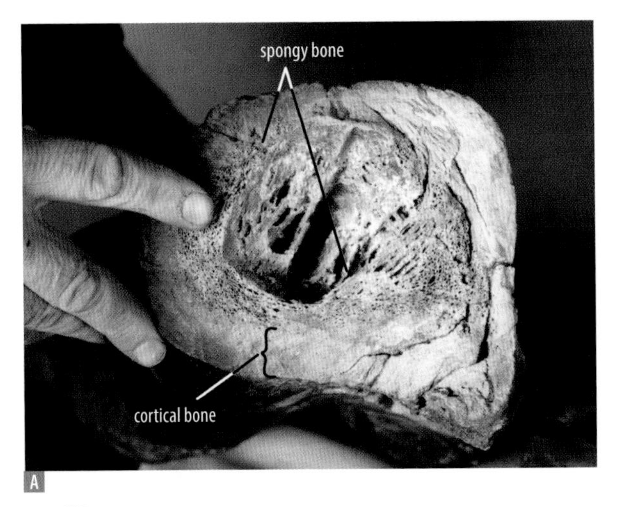

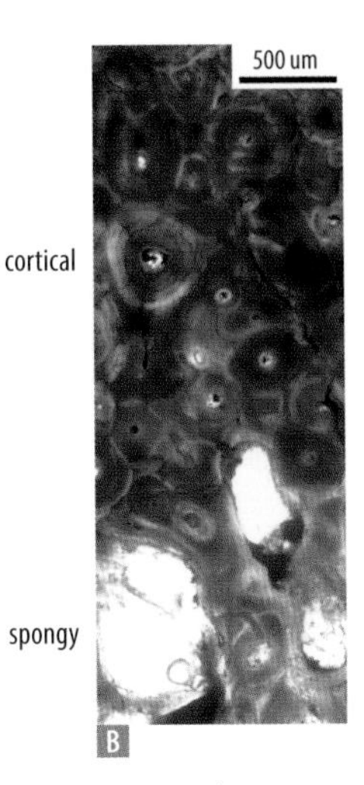

3.6 **A** Bone is not solid, as can be seen in this broken bone of Fran. The large cavity in the center was once filled with marrow. The porous bone is called spongy bone because its mesh-like structure makes it look like a sponge. The denser, outer bone is called cortical bone. **B** Cortical bone is also porous, as shown under a microscope. By slicing and grinding a bone until it is thin enough for light to pass through it, one can see the fine details. *Panel A courtesy of the Black Hills Institute. Panel B is from Michael D. D'Emic, Keegan M. Melstrom, and Drew R. Eddy (2012), "Paleobiology and Geographic Range of the Large-Bodied Cretaceous Theropod Dinosaur Acrocanthosaurus atokensis," Palaeogeography, Palaeoclimatology, Palaeoecology, Vol. 333–34, pages 13–23, with permission from Elsevier.*

bone by calcium carbonate. By the end of the experiment, at ten months, all the bones had a thick rind of cemented sand. But was the calcium carbonate also deposited within the bone? I made a thin section of the bone by cutting a thin sliver with a very thin diamond saw and put it onto a microscope slide. Next I used various grades of sandpaper to make the bone so thin that light could pass through it. Then I used special stains to highlight any calcium carbonate that was present. Under the microscope I saw stained calcium carbonate within the bone where it had not been previously. It was deposited there by bacteria that were feeding on the tissue within the bone (bone marrow, blood, etc.). Their carbon dioxide "poop" had attached to the dissolved calcium in the water to form calcium carbonate on and in the bone. I had artificially fossilized a bone!

The minerals were able to collect inside the bone because bone is not completely solid. Instead, it typically has an inner mesh of bone for the marrow (called spongy bone because it looks like a sponge) and, through the outer layers, or cortical bone, a network of finer tunnels that can best be seen through a microscope (fig. 3.6). In living bone, the walls of these structures are lined with tissue and circulate blood and nutrients to the bone cells. To see for yourself just how much organic material there is in a bone, leave a well-cleaned chicken bone in a jar of vinegar for a week. Then carefully pour it out into your hand over a sink. You'll find that you can actually tie the bone in a knot. What has happened is that the vinegar, a weak acid, dissolved the mineral part of the bone (hydroxyapatite), leaving the organic part, mostly collagen, in the shape of the original bone. To get a real appreciation for just how much organic material there is in a bone, cut up this "ghost" of a bone—you'll be amazed just how much organic material there is.

All of this organic material is potential food for bacteria, which invade bone through the same small openings through which blood vessels enter the bone. If the bone is buried in the presence of groundwater, bacteria can precipitate calcium carbonate as they eat their way through the bone. It was these calcium carbonate deposits that I found within the bone pieces in my experiment. To be absolutely sure that it was bacteria that had deposited the calcium carbonate, I repeated the experiment but added some bleach to keep the water, sand, and bones sterile. This time, no calcium carbonate was deposited on or in the bones, showing that bacteria do indeed play an important role in the fossilization of bone. This experiment has since been repeated by others, including Joseph Daniel and Karen Chin (2010) at the University of Colorado, so my experiment was not a fluke.

Calcium carbonate is not the only mineral in the fossilized Acro bones. I mentioned earlier than pyrite was present outside and inside the bone. This mineral, known as fool's gold, form from the combination of atoms of iron and sulfur (FeS_2), which laboratory experiments have shown can be brought together by sulfur-feeding bacteria. For such bacteria, oxygen is a toxic poison. Not surprisingly, then, bogs and swampy soils are common places for these bacteria to live in and for pyrite to form. In such settings, the abundance of decaying plant material liberates the organic sulfur from the cells, and the water-saturated soil keeps out oxygen. Some of the sulfur and other organic compounds that are released can combine with water to form acids, including sulfuric acid. If these acids concentrated enough, they can completely dissolve bone. That is why it is rare to find fossilized bones in sedimentary rocks rich with plant remains, such as lignite, a low-grade coal. In Fran's case, seasonal floods must have flushed the acids out of the pond periodically to keep the concentration low enough that the skeleton did not dissolve. Still, the environment did not prevent pyrite from forming on and in the bones (fig. 3.7). The pattern of the pyrite suggests it formed

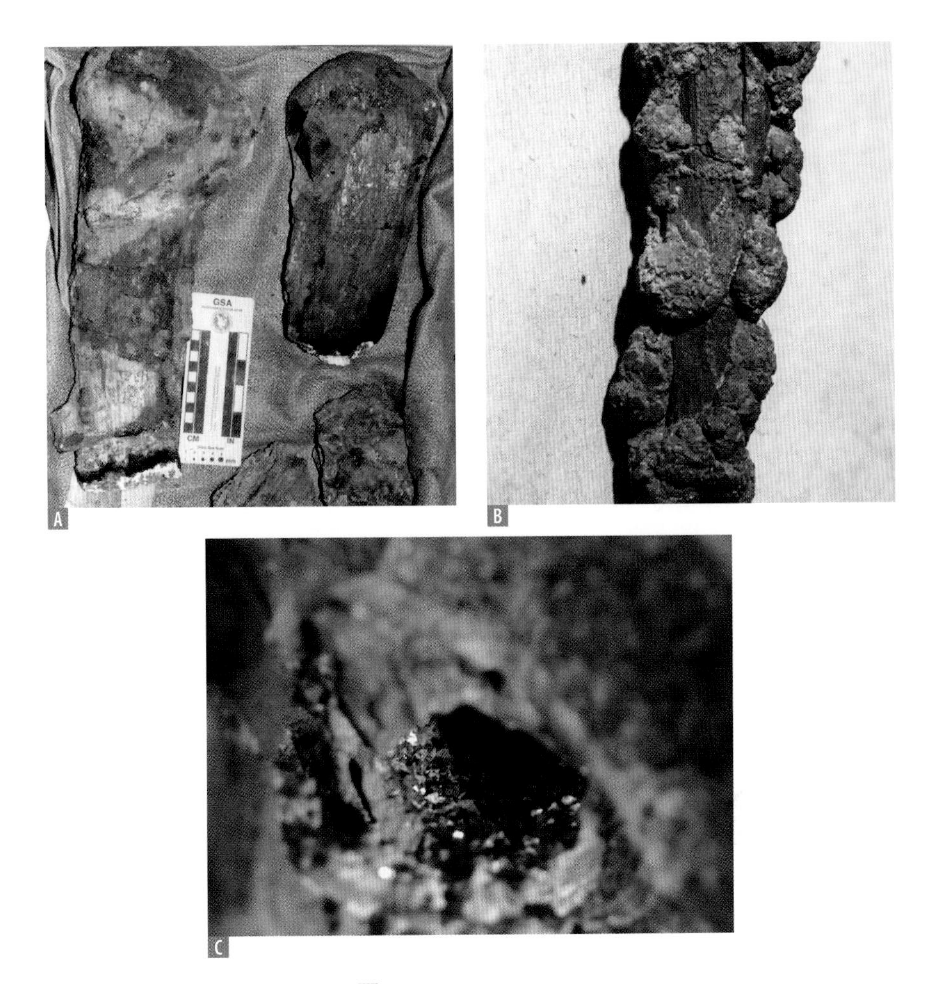

3.7 Minerals on the bones of *Fran.* **A** Calcium carbonate on the spine of a vertebra (left) compared with one after the mineral had been removed (right). **B** Lumps of pyrite on a rib. These lumps were common and difficult to remove. **C** Pyrite crystals lining the marrow cavity of a bone. Pyrite forms from iron dissolved in the groundwater and sulfur released from the decay of organic material in and around the bone. *Courtesy of the Black Hills Institute.*

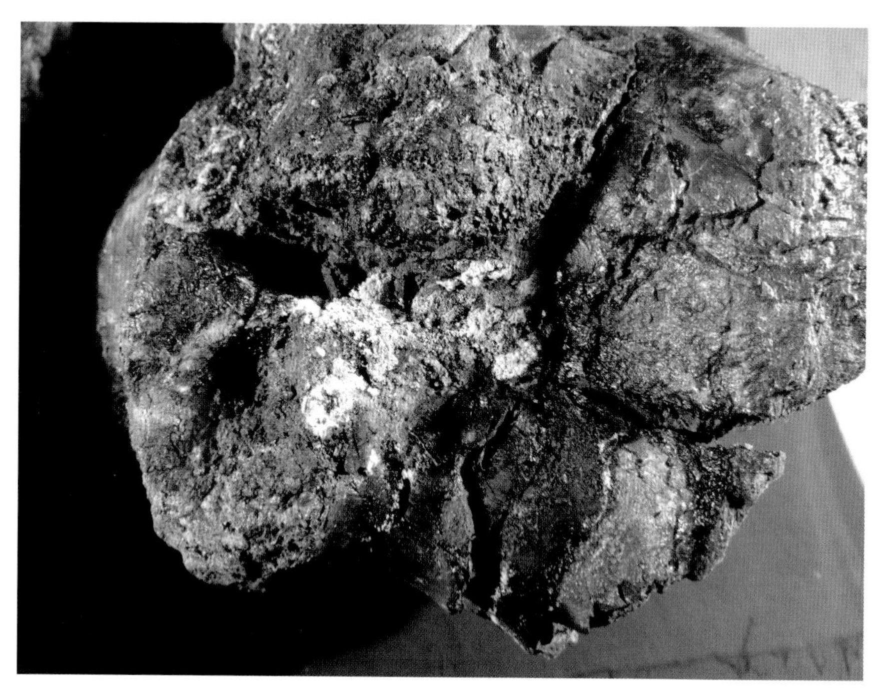

3.8 Breakdown of the mineral pyrite in the presence of a high-humidity environment and oxygen results in a white, frothy powder that pushes fossil bone apart along fractures. Unless halted, this slow-motion "explosion" could reduce this Fran vertebra to a mound of sulfur-smelling dust.

where some soft tissues, such as muscle and gristle, had been present when the bones were buried. This soft tissue, plus any plant material buried with the bones, was the source of the sulfur that the bacteria used to form the pyrite. Removal of this pyrite on Fran was a difficult and time-consuming project in the preparation lab of the Black Hills Institute.

Bacterially created pyrite poses a problem for paleontologists because the mineral tends to break down in humid climates, a condition nicknamed "pyrite disease." When that happens, sulfur is released, which does three things (Shinya and Bergwall 2007). First, the atoms of sulfur combine with water vapor in the air to produce sulfuric acid. This acid attacks the bone, causing further deterioration. Second, a frothy, white iron sulfate powder forms and, as it expands, pushes the bone apart in a slow-motion "explosion" (fig. 3.8). Third, the sulfur also attracts *Thiobacillus* (theo-bah-sill-us), a bacterium that uses sulfur as food, also causing further deterioration. In the 1800s, when pyrite destruction in museum collections was first recognized, fossil bones with pyrite were impregnated with molten beeswax. The wax would seal the bone from oxygen and humidity, thus preventing the pyrite from breaking down. Today, museums store pyrite-rich fossil bones in low-oxygen, sealed containers in cabinets with containers of silica gel (the same granules that are in those little packets that say "Do not eat!" in some packaged foods). Unfortunately, this storage is not a cure for pyrite disease, but it does keep the pyrite from breaking down. The high pyrite content of the bones of Fran is a problem, and the bones will always require special, long-term care to prevent the mineral from breaking down and destroying the fossils.

CHAPTER 4

The Skeleton of *Acrocanthosaurus*

Acrocanthosaurus gets its name from the high spines on the vertebrae, the tallest of which are 2 feet (61 cm) tall! These are some of the tallest spines among dinosaurs, except for those in the spinosaurids, a group of slender-snouted, mostly fish-eating carnivores that have a "sail" on their backs.

We do not have a single complete skeleton of *Acrocanthosaurus*, so how do paleontologists know what its skeleton looked like? Fortunately, the four major partial skeletons have enough of the same bones to show that they all belong to the same dinosaur. Most importantly, though, where a bone is missing in one specimen, in most cases it is present in another (fig. 4.1). So, by using all of the skeletons, I can construct what the skeleton of a single Acro would look like. Although the skeletons are of different-size individuals, I scaled the bones to a single standard size based on the length of the thigh bone, or femur, of Fran. Dinosaur paleontologists typically use this bone in a wide variety of studies because it usually scales with body size or weight. In Fran's case, the femur also came with the most complete skeleton.

The ends of Fran's femur were slightly damaged, but I was able to estimate that it was 50.25 inches (127.6 cm) long. In contrast, the femur of original skeleton described by Stovall and Langston is 45.25 inches (115 cm), which is 10 percent smaller. Consequently, in order to use any of the original skeleton's bones to fill in the missing or incomplete bones of Fran, I needed to enlarge the bones of the original skeleton by 10 percent. Using this scaling technique,

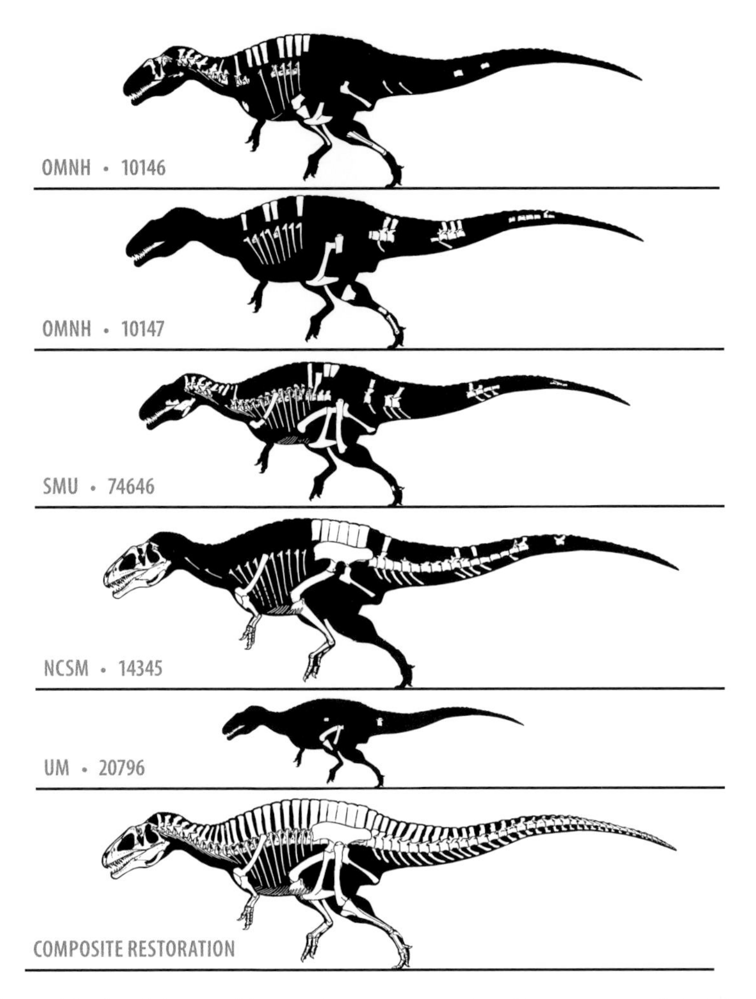

4.1 None of the five skeletons of *Acrocanthosaurus* is complete. However, a complete skeleton can be compiled by using parts from the various skeletons sized to the same scale. This technique is how paleontologists restore many dinosaur skeletons.

I filled out most of the skeleton using the bones of all the skeletons (fig. 4.1). Nevertheless, I was still left with some bones that were missing. For example, not all the ribs are known, but it's a safe bet that the missing ribs look like those that are known, even though the lengths might be different. In addition, I used mirror imaging of bones known from the left or right side of the skeleton to account for missing bones on the other side.

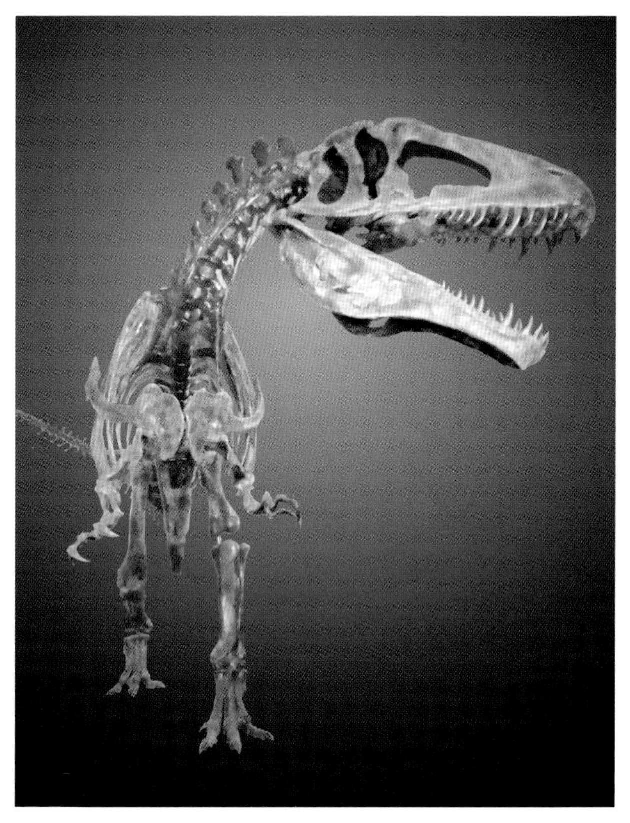

4.2 Cast of Fran's reconstructed skeleton on display at the Museum of the Red River.

Other missing information can be predicted based on other carnivorous dinosaurs, starting with those that are most closely related to *Acrocanthosaurus*: *Carcharodontosaurus* and *Giganotosaurus* first, and then the more distant *Allosaurus*. The skeletons of the two closest relatives are not completely known, either, but *Allosaurus* is known from several nearly complete skeletons. For instance, because *Allosaurus* had about forty-five vertebrae in its tail, I can assume that *Acrocanthosaurus* did, too. Using this reasoning and scaling a skeleton of *Allosaurus* to Fran's size allowed me to reconstruct the missing tail and some of the toe bones to reconstruct a complete skeleton of *Acrocanthosaurus*. The Black Hills Institute used all of these techniques to reconstruct Fran for display (fig. 4.2). We will know how accurate this reconstruction is only when a complete skeleton is found.

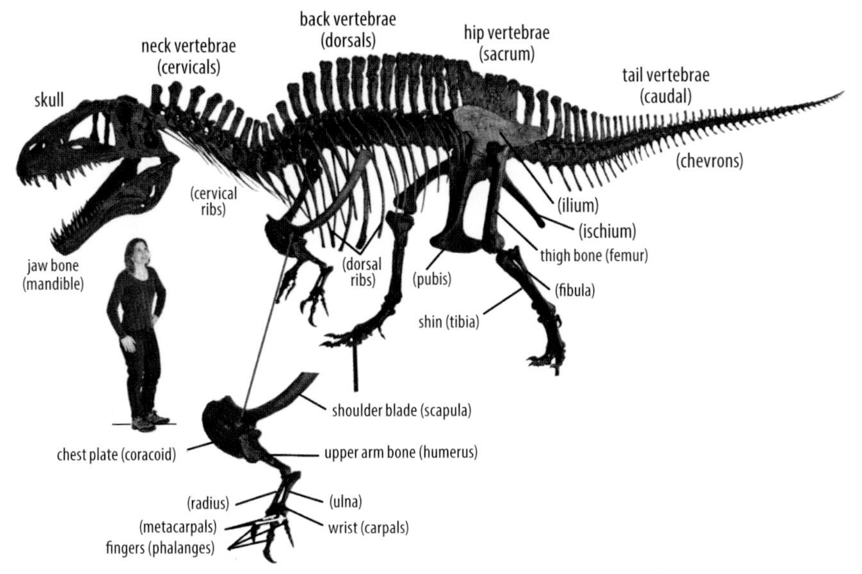

4.3 Common and scientific names (in parentheses) of the major bones of *Acrocanthosaurus*. A human is shown for scale, and the bones of the forelimb are shown enlarged. This guide should help you follow the discussion of this chapter. *Skeleton image courtesy of the Black Hills Institute.*

Before we look at the skeleton more closely, let me answer a question that I'm frequently asked: how many bones did *Acrocanthosaurus* have? It depends. A hatchling probably had a total of 608 bones—136 skull bones, 358 bones along the middle of the skeleton (vertebrae, ribs, etc.), and 114 limb bones (shoulder blade, arms, pelvis, etc.). Some of these bones, such as those that make the braincase and the vertebrae, fuse together with age. Fran, for example, probably had 451 bones: 123 bones in the skull, 216 bones along the middle of the skeleton, and 112 bones in the limbs. Now let's look more closely at some of the individual bones to get a better understanding of this dinosaur (fig. 4.3).

THE SKULL

Langston had only pieces of the skull when he wrote his master's thesis (1947). He attempted a rough (very rough, actually) approximation of the complete skull (fig. 4.4A). Despite stating that *Acrocanthosaurus* was a close relative of *Allosaurus*, he did not use the *Allosaurus* skull as a guide. If he had, the result

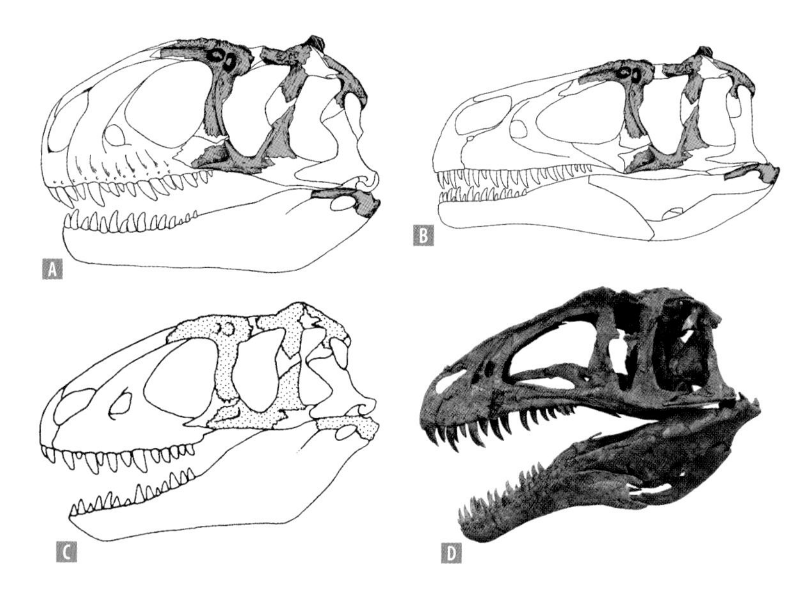

4.4 A Langston's (1947) reconstruction of the skull of *Acrocanthosaurus* showing the skull bones available to him. Langston never gave a reason for making its snout so big and deep. B The result if Langston had used a skull of *Allosaurus*, the closest relative of Acro known at the time, as a guide. C Langston's (1974) revised skull, in which the deep snout was revised and made to be more tapered. This version is closer to D the actual skull of Fran found two decades later.

would have been closer to the actual skull found with the skeleton of Fran (compare fig. 4.4B with fig. 4.4D); a version he did at a later date was a little more accurate (fig. 4.4C).

When Fran's skull was found, it was flattened from side to side by the pressure of the overlying rock. It was carefully taken apart at the Black Hills Institute, and each piece was molded with silicone rubber and cast in plastic. The cast parts were then pieced back together to reconstruct the skull in an uncrushed state. The result was a skull 56.5 inches (143.5 cm) long, almost the size of a *Tyrannosaurus* skull (58 inches, or 147.3 cm). But the skull of *Tyrannosaurus* is built like a bulldog's, whereas the slender skull of Acro is more like a greyhound's. This difference has much to do with diet, as revealed by the teeth. Those of *Acrocanthosaurus* are slender blades, which are well adapted for slicing through meat, whereas the banana-shaped teeth of *Tyrannosaurus* were adapted for crunching through bone (but not gnawing like a dog). Bone crunching requires brute force, and that requires a lot of strong jaw muscles. To accommodate those muscles, the rear of the skull in *Tyrannosaurus* widened to make more room for the jaw-closing adductor muscles. Acro lacks this expansion, so it must not have had such powerful jaw muscles. Still, its 3.75-inch (9.5-cm) serrated teeth could slice through the hide and muscle ("meat") of most dinosaurs. We'll look at the teeth in greater detail below.

The skull of Acro has five major openings on each side (fig. 4.5). At the front of the snout is the long nostril opening; this opens into a large chamber that fills most of the snout, where tissue for smell was located. Next is the large, triangular, antorbital fenestra, which means "window in front of the eye socket." (Many people mistakenly think that this opening is the eye socket.) Studies of birds show that this "window" in Acro housed part of a widespread, skull-lightening sinus system. The eye socket (orbit) is a keyhole-shaped opening that is partly divided by a triangular piece of bone. The eyeball sat in the upper, rounder part of opening. We'll look at eyeball size in more detail in chapter 6. At the rear of the skull on each side is a tall, oblong opening called the lateral temporal fenestra ("side window in the temples") and a smaller counterpart on top the skull called the supratemporal fenestra ("upper window in the temples"). These two "windows," especially the lateral temporal fenestra, may have given the jaw-closing adductor muscles room to bulge out when the jaws bit down hard. You can better understand this by lightly pressing your finger tips on your temples and biting down. You'll feel the muscles bulge a little.

The major bones of the skull, such as the paired frontal bones on top the skull, have interlocking joints, or sutures, which look like tiny interlocked fingers of folded hands. Such sutures rigidly lock certain bones together. Other sutures—such as the one between the two bones that form the rear of the eye socket, called the postorbital and jugal bones—have an overlapping suture that allows some movement between the bones. A similar feature is seen in many birds, especially between the braincase and beak, and is called kinesis, which means "movement." At one time it was thought that kinesis in carnivorous dinosaurs, such as Acro, allowed them to open their jaws apart sideways, like a snake does. However, more recent studies (e.g., Rayfield 2005) of kinesis show that this ability for some bones is located in regions of the skull that would experience the greatest stresses during a bite. Kinesis in Acro, then, acted as a shock absorber to reduce the stress on the skull when the animal bit down on prey. One of these kinetic bones near the back of the skull is indirectly involved with the lower jaw. Called

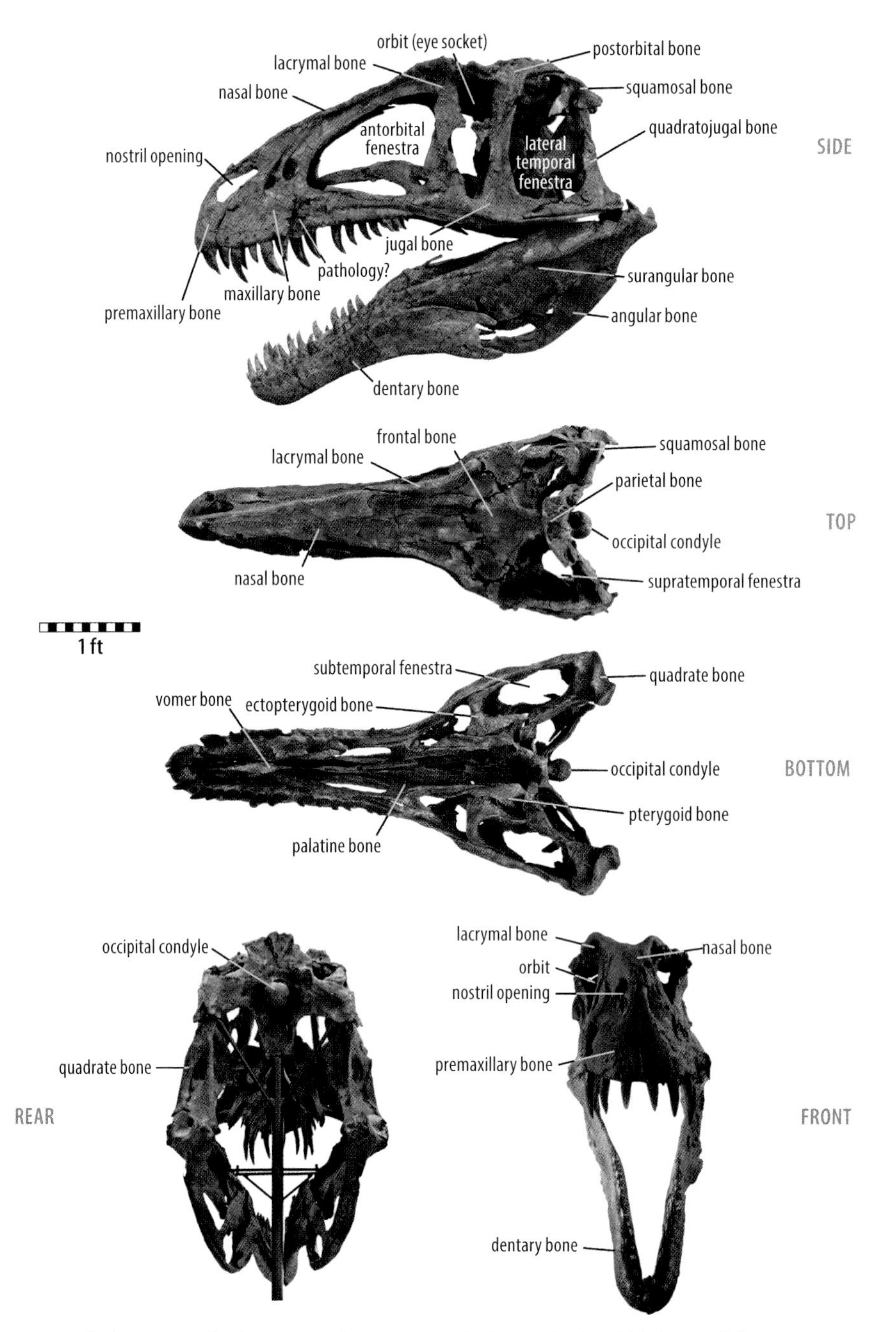

4.5 Skull of Fran in multiple views with parts named. This guide should help you follow the discussion in this chapter. Also note the possible injury that Fran sustained during life marked "pathology?" on the maxillary bone. This possible pathology is discussed in chapter 7. *Images courtesy of the Black Hills Institute.*

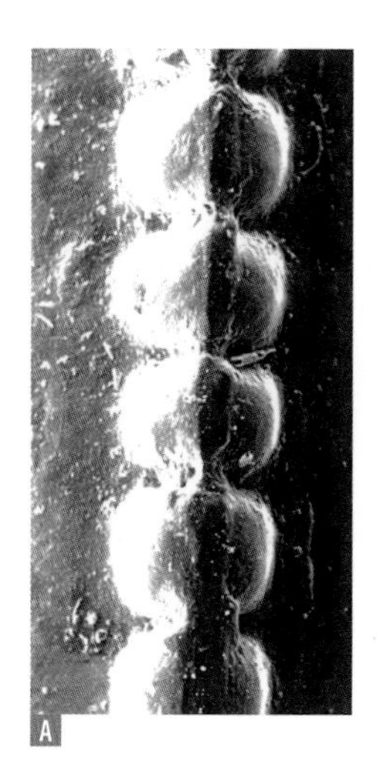

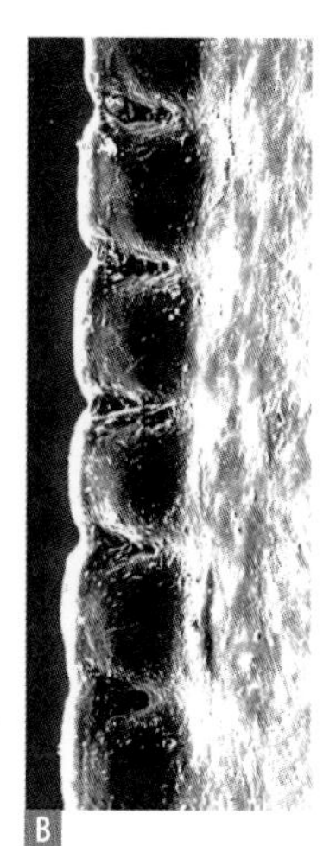

4.6 Scanning electron microscope close-up showing the serrations of an *Acrocanthosaurus* tooth (SMU 74646) in A cutting-edge view and B left side view. These serrations give the teeth the jagged edges to cut through strands of muscle fiber. *Courtesy of Jerry Harris.*

the squamosal, it has a socket for the top part, or head, of the quadrate bone, to which the lower jaw attaches. Langston (1947) noted that within the socket of one of his specimens there was some bone growth, which indicates a mild bone infection. He suggested that had the animal lived longer, movement between the squamosal and quadrate bones would have been restricted. How this infection might have affected eating is not known, but it was probably painful when the force of the bite was transferred up the quadrate.

The upper and lower teeth are fat blades with sharp edges on both the front and back edges. The edges have tiny serrations, which make the teeth look and act like saw blades or steak knives (fig. 4.6). The number, sizes, shapes, and positions of these serrations differ among carnivorous dinosaurs; Acro has about fifty-five serrations per inch along the back edge, whereas *Tyrannosaurus* has forty-five. Why is this? Perhaps we can learn a lesson from hacksaw blades, which have different numbers of teeth (serrations) per inch, depending on the material to be cut: the fewer serrations per inch, the harder the material that can be cut. If this holds true for dinosaur teeth as well, then Acro must have bitten softer material, such as skin and muscle, but not bone. Even so, Acro probably had a powerful slicing bite.

Like most dinosaurs, Acro replaced its teeth on a regular cycle throughout its life. Most mammals—dogs, cats, humans, and so forth—have only two sets of teeth (baby teeth and adult teeth) during their lives, whereas dinosaurs, including Acro, replaced their teeth about every fourteen to twenty-four months (Erickson 1996). A thirty-year-old Acro would have completely

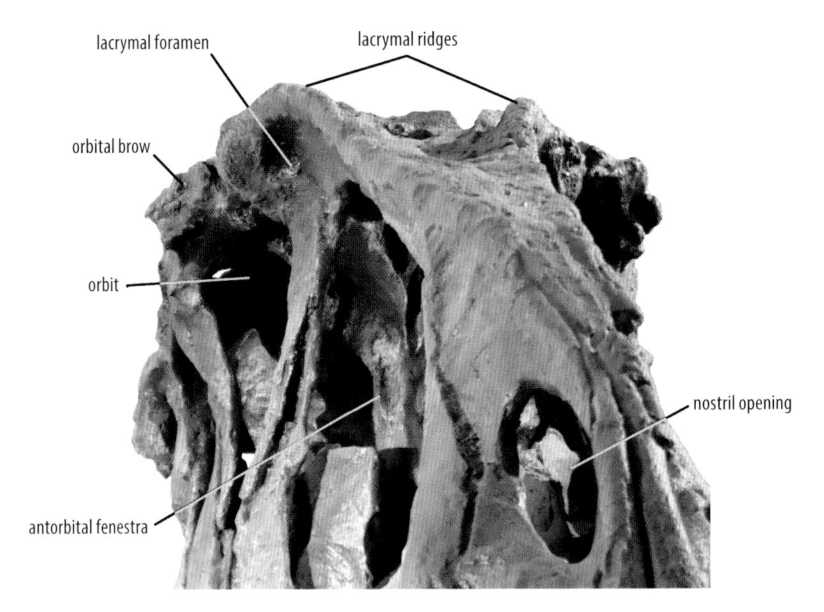

4.7 Front view of the skull of Acro with some of the landmarks named. This view shows the paired ridges developed from the lacrymal bones in front of the eye socket. The lacrymal foramen is an opening into a sinus cavity within the bone. The orbital brow is a shelf above the eye socket that may have made the eyeball appear to be sunken into the skull, giving the animal a "beady-eyed" look.

replaced all its teeth about fifteen to twenty-five times. At any point during Acro's life, several replacement teeth were forming below the gum line. These replacement teeth pushed upward (or downward for the upper teeth) at about the same rate that the erupted teeth wore down. In carnivorous dinosaurs, this wear was caused mostly by the tooth striking bone during feeding. A tooth's serrations will become dull with use, as a saw blade does, and sometimes the tip of the tooth shattered if it struck bone at just the wrong angle. Replacement was therefore important and began with a new tooth forming in miniature at the bottom of the tooth socket. Gradually the replacement tooth became larger and pressed against the root of the tooth it would eventually replace. This pressure triggered the bone cells in the root to erode the root by removing the bone calcium. This erosion weakened the tooth so that it would eventually fall out, which apparently happened mostly during feeding because of the stresses placed on the tooth. That is why isolated theropod tooth crowns are commonly found among dinosaur bones. The replacement tooth continued its slow growth upward in the socket until the crown fully erupted. Eventually, it too would wear and be replaced. The replacement cycle happened in alternating waves along the tooth row so that no large gaps ever occurred. That is why the teeth in a theropod's skull, as seen in side view, are of different lengths (fig. 4.5).

Along the top of the snout is a pair of ridges formed by the lacrymal bones that extend to a thick brow above the eye socket (fig. 4.7). These are most noticeable in front view. Most carnivorous dinosaurs have these structures, although in some, such as *Allosaurus*, they form a short, triangular horn. The lacrymal bone of Acro is pierced by a small opening, the lacrymal foramen, which leads to a cavity that is part of the sinus system of the antorbital fenestra. Above the eye socket is the shelf-like orbital brow. Unless the eyeballs bulged out like they do in a deer, the eyes would have been sunken, giving Acro a beady-eyed look. Although they faced mostly to the sides, the eye sockets are visible in front view, which means that Acro could look straight ahead and see in three dimensions (more about this in chapter 6), which is an ability shared by only a few other carnivorous dinosaurs. This capability is important for judging distance and is especially important for predators when catching prey. However, to get a good look at something, Acro probably cocked its head so that the eye on one side was facing toward the object, the same way that birds do today.

At the back of the skull, there is a half sphere (hemisphere) just below the opening into the braincase. This half sphere, called the occipital condyle, is where skull attached to the neck in a ball-and-socket joint, which allowed the skull to pivot on the neck.

THE REST OF THE SKELETON

The neck bones, called cervical vertebrae, have tall spines (neural spines) to which muscles, tendons, and ligaments attached (see fig. 2.1 in chapter 2). Ligaments are made of a tough, stretchable tissue that connects one bone to another, and they often occur as thin sheets. Tendons, on the other hand, connect muscle to bone; for instance, the Achilles tendon on the back of your foot connects the calf muscle, or gastrocnemius, to the heel. Both tendons and ligaments can leave marks, called scars, on the bone where they attached. These scars appear on the spines of Acro's neck vertebrae, indicating that Acro was unusual among carnivorous dinosaurs in that it had a very powerful neck. These muscles were used to pull the head back when Acro ripped flesh from its meal. The vertebrae of the back also have very tall spines, but these get progressively taller from the base of the neck to the middle of the back; the tallest of these is almost 2 feet (61 cm) tall! The tall spines continue to the vertebrae of the hip (sacral vertebrae) and begin to get lower on the tail vertebrae (caudal vertebrae). Because the purpose for these tall spines is so controversial among paleontologists, they need their own chapter (chapter 5). Below the spines, each vertebrae has a spool-shaped part called a centrum. This part is pierced by small openings that lead into cavities inside the bone. In life, these cavities were indirectly connected to Acro's breathing system but were not active parts of it (see chapter 6).

The shoulder blade, or scapula, is a long, slightly curved blade of bone that provided a large attachment surface for the muscles that pull the arm backward. At its front, the shoulder blade curves across the chest, where it contacts a plate of bone called the coracoid. This plate of bone provided a surface for the attachment of muscles that connect to the upper arm bone, called the humerus, and pull the arm inward toward the chest. The lower arm has two bones, the radius and ulna. In humans, the radius can roll over the ulna so that the palms can face up or down. In all dinosaurs, however, including Acro, the radius and ulna have a flat surface between them at the elbow that prevents the radius from rolling over the ulna. That means the palms could not face downward, as is often illustrated for carnivorous dinosaurs; rather, the palms faced inward, toward each other, so that the hands were always ready to grab. As dinosaur paleontologist Jerry Harris tells it, theropods could hold a basketball, but they could not dribble.

The wrist is made of disk-shaped bones, or carpals, that allowed the hand to move. One of these wrist bones is shaped like a half-moon and therefore is called a semilunate carpal. It allows some movement of the hand in an up-down direction, a movement that humans do not have (equivalent to facing your palms toward each other and being able to bend your wrists so far downward that your pinkies can touch the side of your arms). In other carnivorous dinosaurs, especially the "raptors" such as *Velociraptor* (vel-oss-er-ap-tore), this up-down ability actually allowed the arms to fold like a bird's wing. Acro, however, could not quite achieve that. The hand of Acro has three fingers, including a thumb that is angled away from the other fingers. The hand bones include the long metacarpals that form the palm and the phalanges, or finger bones. Each of the fingers ends in a curved claw for holding prey.

Acro's arms are almost as long as a human's arm, or the arm of a *Tyrannosaurus*, which is far too short to have allowed Acro to walk on all fours.

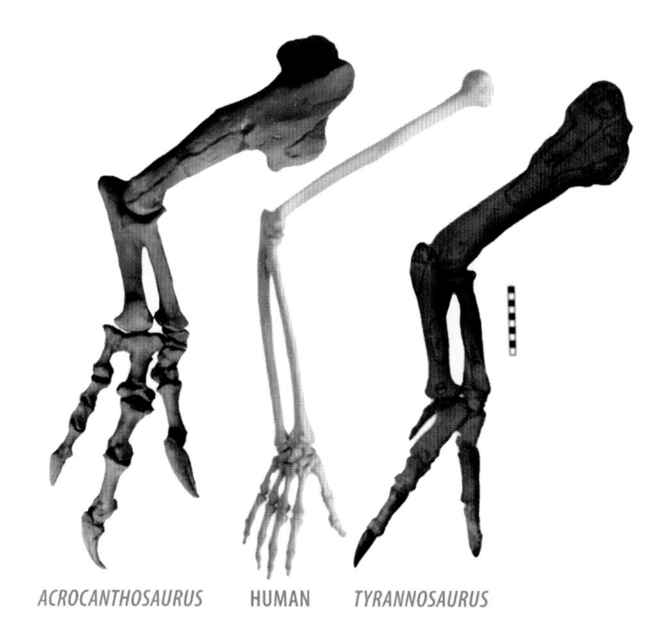

4.8 Comparison of the arm bones of *Acrocanthosaurus*, a human, and *Tyrannosaurus* to scale. Note how skinny the human bones are in comparison with the two dinosaurs'. The heftiness of the dinosaur bones and large scars for muscles show that these arms were powerful. Scale is 10 cm.

However, although the arm is as long as a human's, the bones are much bigger in diameter and thus much stronger (fig. 4.8). In addition, the bones have very large projections to which muscles attached. The result is that the arm of Acro was very powerful. However, a study by paleontologists Phil Senter and James Robins (2005) of the movement possible at the shoulder and elbow of Acro shows a surprisingly limited range of motion. For instance, the hands could not reach past the base of the neck, meaning that the hands could not reach the mouth. Instead, most motion was toward the chest, which makes sense if the arms were used to hold struggling prey in a death embrace while the teeth and jaws did all the killing work (Carpenter 2002).

On each side of the pelvis are three bones that connect to form the hip socket. The upper bone is a long plate called the ilium. It had large muscles that attached to the upper leg bone, the femur, for moving the leg back and forth. Those that attached to the front of the ilium and hind leg pulled the leg forward, whereas those that attached on the back side pulled the leg back and, in doing so, moved the body forward. Because it takes more muscle to move the body forward than it takes to lift and move the leg forward, those muscles on the back of the leg are larger. (You can see this for yourself the next time you eat a chicken thigh.) Below the ilium is a long bone called the pubis, which projects down and forward, and another rod-shaped bone, the ischium, that projects down and backward. The shaft of the pubis provided an attachment surface for a few thin muscles that pulled the leg inward so that the body weight was kept over the hind legs. The pubis ends in a boot-shaped structure, or foot, that supported the animal when it squatted to rest. The impressions made by the pubis in a squatting carnivorous dinosaur have been described from Lower Jurassic rocks in Massachusetts (Lull 1953) and Utah (Milner et al. 2009). The ischium had muscles that pulled the legs back and inward.

Acro had long legs made up of the thigh bone (femur), the shin bone (the large, inner tibia and the slender, outer fibula), the ankle (astragalus and calcaneum), and the foot (metatarsals and phalanges). At its upper end, the femur has a short cylindrical part, called the head, which projects about 90° to the long, main part of shaft of the bone; the head inserts into the hip socket. In crocodilians and lizards, the femur head angles upward and inward at about 60°, allowing the leg to sprawl when the reptile is basking. Birds, however, have an in-turned head set at 90°, similar to that of Acro. This places the legs beneath the body when walking or running. Birds may squat to rest or brood eggs, but they can't sprawl. Because Acro had legs like a bird's, not like those of a crocodile or lizard, it most certainly walked like a bird and squatted rather than sprawled when resting.

I mentioned earlier that we can reconstruct the incomplete tail of Acro by using *Allosaurus* as a model. If the result is anywhere close to accurate, it would suggest Acro had a long tail to counterbalance the front of the body over the hind legs. The tail vertebrae of most theropod dinosaurs tightly interlock, which prevents them from using the tail like a whip or curling it; instead, the tail was held straight out behind the horizontal body as a counterbalance to the front of the body over the hind legs. Most of the tail's movement happened near the hips, which would allow the tail to swing a few feet left to right like a pendulum; this kept the animal's center of gravity over the weight-bearing foot as it walked or ran.

CHAPTER 5

Why Tall Back Spines?

The tall neural spines on the neck-, back-, and tailbones of *Acrocanthosaurus* deserve their own chapter because there is much to say about them. Acro is one of about a dozen dinosaurs, mostly from the Cretaceous, that had tall neural spines. Langston (1947), the first to ponder why Acro had such tall spines, suggested that the spines were embedded in muscle, much like those in the shoulder region of the American bison (commonly known as a buffalo). He dismissed the possibility that the spines were simply covered with skin to form a "sail" or ridge, saying the notion was "certainly unacceptable." He also considered and dismissed a mechanical explanation, such as the spines making a deep, narrow tail for crocodilian-like swimming or acting as anchors for large neck and back muscles to support the large head. The problem with these ideas, Langston noted, is that the tail was rather stiff, so it was not well adapted for propelling the animal in water. Furthermore, he noted that the skull of Acro was not proportionally larger than that of *Tyrannosaurus*, which lacks tall spines for large muscles, meaning that especially large muscles were not needed to support a head that size. Langston concluded that the tall spines might be a "sexual characteristic," although he never clarified whether he meant they indicated masculinity in males or femininity in females.

More recently, Jack Bailey (1997) elaborated on Langston's "buffalo-like" hump idea (fig. 5.1). In buffalos, the shoulder hump is made of muscle and fat and is well developed in the adults. Bailey suggested that a similar hump occurred along the back and tail of Acro, but this idea or hypothesis has never gained much acceptance among dinosaur paleontologists because most of us

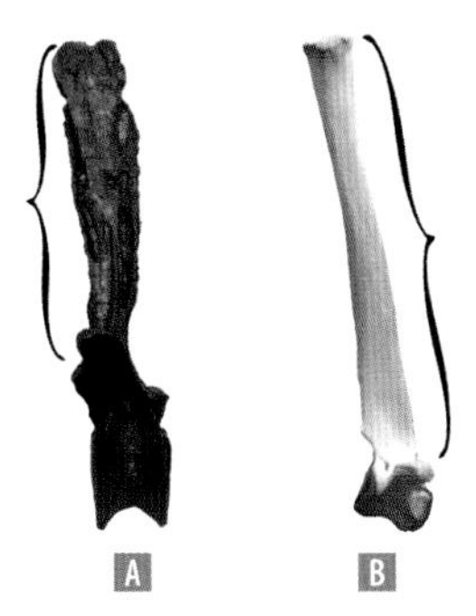

5.1 Comparison of A Fran's tall spine (marked by brackets) with B that of a modern American bison. These are not to scale: the bison vertebra is less than half the size of the Acro vertebra.

think the similarities in the tall spines served a different purpose. In male buffalo the large foreleg muscles of the shoulder power their forward lunge in head-to-head butting. (This butting is why they also have a thick mat of fur padding, like a football helmet, on their broad foreheads.) In *Acrocanthosaurus* the skull was narrow and the bones much too thin to hold up to head butting, so it is doubtful they did that. Furthermore, the tall spines in Acro are not limited to the shoulders but extend from the neck to at least the front half of the tail. So whatever the tall spines were for, we need to look elsewhere for the answer.

We do know that one important purpose for the spines of vertebrae in other animals, from fishes to humans, is the attachment of back muscles. These muscles not only support the back but also bend the back from side to side during walking. The big question then becomes just how much back muscle Acro had or needed. If they had more spine than that needed for muscle, what was the use of the excess spine? Let's look at the pros and cons of the three most likely hypotheses (fig. 5.2).

TALL SPINES FOR BACK MUSCLES

Unlike arm or leg muscles, which start on one bone, cross a joint, and attach at another bone, back muscles have a very complicated arrangement (fig. 5.3). Some muscles split apart into numerous branches, or numerous branches come together to form a single muscle. The branches originate on separate vertebral spines or on the sideways projection on the vertebra (called transverse processes) to which the ribs attach. These muscles fill the space between the spine and the rib. Some of the best cuts of beef come from this region, including T-bone and rib-eye steak. For Acro that would mean a huge steak, with an area of 95 square inches (613 square cm), which is more than

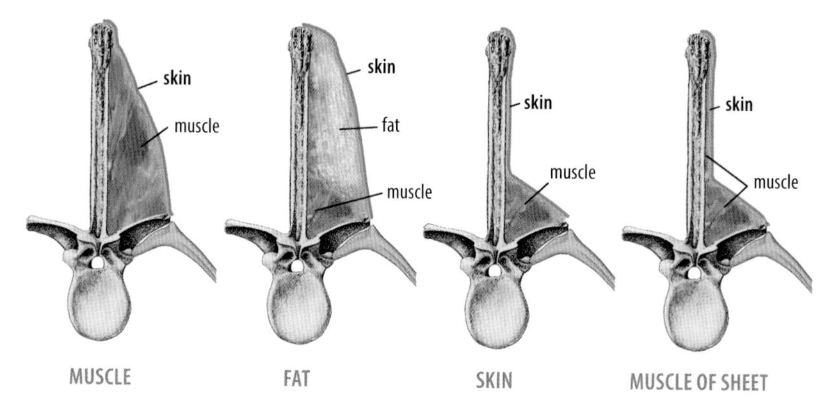

5.2 The first three panels illustrate the three major models for explaining the high spines of *Acrocanthosaurus*, proposing (from left to right) a large mass of muscle (strong back), fat storage (protection from heat or for lean months with no food), and skin over the spine (body temperature control). The presence of ridges and rough patches indicate that some muscle was present, though probably not the mass envisioned by Bailey (1997). An alternative suggestion, shown in the far right panel, proposes thin sheets of muscle along the spine.

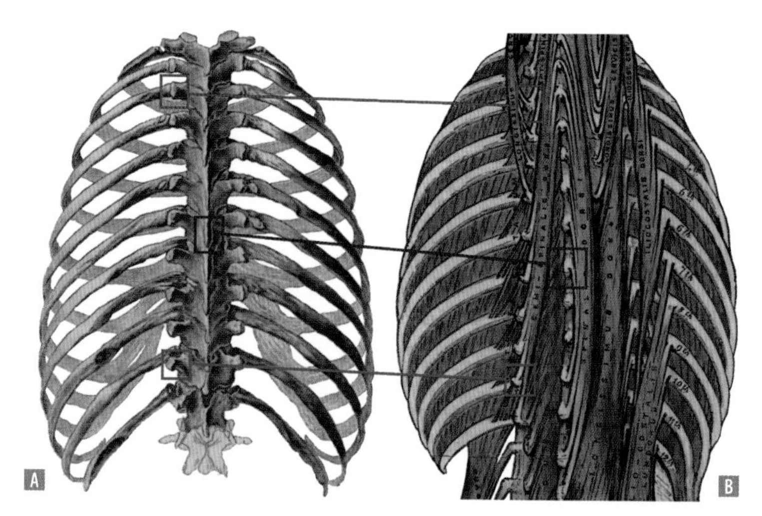

5.3 A Vertebrae of the human back and B the muscles that span and connect numerous vertebrae. When the muscles on one side contract, they pull the back in a curve. In a walking *Acrocanthosaurus*, the alternating contraction of muscles on opposite sides of the back would curve the back from side to side, keeping the body centered over the foot. The colored boxes show the same feature or landmark to make it easier to compare the muscled back (B) with the skeleton (A). *Images from Gray and Lewis (1918).*

four rib-eye steaks! (I'd bet Acro tasted like chicken because Acro is closely related to both the chicken and alligator; alligators taste like chicken, too.)

At 39 feet (12 m) long, Fran is about the size of a *Tyrannosaurus*, but the vertebral spines in *Tyrannosaurus* are nowhere near as exaggerated. Proportionally, the spines on *Tyrannosaurus* vertebrae are no larger than they are in most theropod dinosaurs, regardless of size; I'll call these proportions the norm. *Tyrannosaurus* did everything that we would expect of a giant predatory dinosaur: it hunted, brought down prey, grabbed with its teeth, and ripped the flesh using its neck and back muscles to pull its skull up and back (more on this in chapter 8). *Acrocanthosaurus* should therefore have been able to do the same with normal, *Tyrannosaurus*-sized back muscles. I could argue that Acro had really weak muscles and needed more of them to get the same results as *Tyrannosaurus*, but for the most part muscle fibers have the same pulling strength, regardless of whether they're in a lizard or an elephant. Stronger muscles are the result of an increase in the number of muscle fibers, which makes the muscles larger in diameter. Another way to think about this is to imagine a ten-person tug-of-war team competing against a five-person team: because there are more people in the larger team, they are both more powerful and take up more space.

If the entire space next to Acro's spine was filled with muscle, it would be way out of proportion with other carnivorous dinosaurs, and having so much extra muscle would serve no conceivable purpose. It would also demand a lot of the body's physiology to create excess muscle that it never needed. Therefore, it makes little sense to me for Acro to have an abnormally powerful neck, back, and tail. I discount this hypothesis.

TALL SPINES FOR FAT STORAGE

Another of Bailey's (1997) suggestions is that the tall spines in Acro supported a large hump of fat. He argued that the fat deposit may have served one or more functions: that it was an energy source to get the animal through times when prey was scarce, that it insulated the body from the high temperatures of its environment (see chapter 9), or that it gave the body enough bulk) that the body temperature would stay more or less constant, similar to how a bathtub full of hot water cools more slowly than a sink of hot water.

Although these are creative ideas, finding support for any of them is impossible because we do not have a living Acro to study. We can, however, point out some objections that cast doubt on the fat hypothesis. Adding weight in the form of fat deposits would undoubtedly slow the predator down, making it an inefficient hunter. Predators today tend to be sleek and have little excess fat. (We won't count pudgy dogs and cats because that condition is not natural.) Another problem with the fat hypothesis is that it is difficult to understand how a strip of fat along the back would insulate the rest of the body from heat. So much of the body would be exposed to the sun that the back fat would be worthless in keeping the animal from overheating. Finally, how much fat would be needed to create the "mass effect," like a bathtub of water? Too little would do no good, and too much would be a waste. Five-ton elephants living in hot, dry environments (as Acro did) do not use this strategy today, so it is hard to understand why Acro, weighing almost the same, would do so. Most animals use behavior to control body temperature, such as basking in the sun to warm up (e.g., lizards, crocodilians, dogs, horses, and elephants) and seeking shade or wallowing in water to cool down. It seems more probable that Acro did the same thing, so I dismiss the fat hypothesis as unlikely as well.

TALL SPINES COVERED WITH SKIN

The other suggestion is that Acro had normal-sized back muscles at the bases of the spines but that most of the spine extended beyond the muscles and was covered with skin. These skin-covered spines would form a tall ridge or sail down its back, but the purpose of such a structure is more difficult to determine. It has been suggested that a sail along the back could help to control body temperature, a process called thermoregulation. This idea is modeled on another Oklahoma native, the sail-backed *Dimetrodon* (die-me-tro-don). This distant, early relative of mammals lived 45 million years *before* the first dinosaur and 167 million years before Acro. Paleontologists Al Romer and Llewellyn Price (1940) suggested that the sail in *Dimetrodon* acted as a heat collector to warm the body by absorbing more sunlight during basking and that it could also shed heat if the body was too warm (see also Tracy, Turner, and Huey 1986). For a "cold-blooded" predator, they reasoned, warming up and becoming active before your prey did could be a great advantage. The presence of a sail in Acro and other high-spined dinosaurs, it could be argued, served a similar body-temperature-control or thermoregulatory function.

This idea presents several problems. First, there is a large range in the proportion of spine height to vertebra size among dinosaurs and other prehistoric animals. In fact, you could almost line them up, starting with their "normal" proportion in *Tyrannosaurus*, then slightly taller in *Allosaurus*, to taller still in *Acrocanthosaurus*, to even taller in the herbivorous *Ouranosaurus* (oo-ran-oh-soar-us), and, finally, to absurdly tall—up to 6 feet (1.8 m) in *Spinosaurus* (spine-oh-soar-us). We might expect to see taller spines in animals living away from the equator in cooler (but not cold) regions, where having a basking sail would warm up the body faster. (In a cold climate, heat loss would be too great to allow any advantage from basking.) Rapid warming of the body could be a great advantage for a "cold-blooded" predator, but what if Acro was "warm-blooded"? In that case, the taller spines might act like a radiator to get rid of extra body heat. Of course, this poses a whole set of other problems. The biggest is that the body temperature would have to be higher than the air temperature so that heat could radiate from the body, thus cooling the animal. That might happen during the rainy season, when air temperatures were cooler, but what would happen during the height of the summer, when air temperatures exceeded 100° Fahrenheit (38° Celsius)? The animal would risk death from absorbing too much heat. One solution to that problem would be to restrict blood flow to the high spines, but then the animal would have no way to get rid of excess heat. These problems suggest to me that these tall spines had nothing to do with controlling body temperature—or, if they did, that it was a secondary function and not critical to the animal's day-to-day living.

This idea of skin over the spines has one serious flaw: as Jerry Harris (1998) has pointed out, ridges, rough patches, and other scars on the spines show that ligament and muscle was present. However, this does not necessarily mean that the muscle mass was large, as Bailey (1997) suggested. In fact, it makes more sense for the muscle to have been in thin sheets because this would allow some side-to-side flexibility to the back, which is necessary for walking on two legs.

In deciding the function of tall spines, we need to keep in mind that Acro and other tall-spined animals were the exceptions in their day, not the rule. They lived among normal-spined animals, some of which were smaller and some of which were much larger. If the tall spines were for thermoregulation, then the normal-spined neighbors would have overheated unless they could thermoregulate using behavior: for instance, by basking when cool and seeking shade when hot. Many animals use behavioral thermoregulation, such as a lizard basking on a rock or a housecat sunning in a window, so having a radiator on the back is not a requirement.

AN ALTERNATIVE SUGGESTION

I feel that one point about sail-backed animals hasn't received enough attention among paleontologists: where they lived in relation to the equator or climate zones. Because the continents move on the surface of the earth, there were times when various regions inhabited by various sail-backed animals were much closer to the equator than they are now. For example, 280 million years ago, the sail-backed amphibian *Platyhystrix* (plat-ee-hiss-tricks), which looked like a giant, sail-backed salamander, waddled around southern Colorado. At the time, the region was only about 5° north of the equator, so its environment was quite hot. This amphibian lived about 45 million years before the first dinosaur and more than 167 million years before Acro. The sail-backed early mammal relatives *Dimetrodon* and *Edaphosaurus* (ee-daff-oh-soar-us) were contemporaries of *Platyhystrix* and lived in the same narrow zone near the equator. If we jump forward in time to 120–95 million years ago, when global temperatures during the Cretaceous were at their peak and the earth's poles were devoid of ice, as demonstrated by the types of fossil plants there, we find a variety of tall-spined and sail-backed dinosaurs living alongside normal-spined dinosaurs in wide regions across the planet. In the region that became northwestern Africa (Morocco, Algeria, Niger, etc.), which was farther south and straddled the equator at that time, there were the sail-backed plant-eater *Ouranosaurus* and the carnivorous *Spinosaurus*. Farther north from the equator were the tall-spined *Acrocanthosaurus* around 32° north and *Baryonyx* (barry-on-icks) and *Becklespinax* (beck-el-sine-ax) around 40° north on warm, tropical islands.

What all these high-spined and sail-backed animals had in common is that they lived at times and places in which the climate was warm to hot (temperatures frequently more than 100°F, or 38°C), so it would seem that basking in the sun was really not necessary. I do not think it is a coincidence that the tallest-spined animals (*Dimetrodon*, *Edaphosaurus*, *Ouranosaurus*, *Spinosaurus*) lived close to the equator and the shorter-spined animals (*Acrocanthosaurus*, *Baryonyx*) lived farther away. The year-round hot weather near the equator allowed the animals to develop tall sails without risk of serious body heat loss, whereas living farther from the equator did not allow for such tall sails. If any of this is true and the sails were not for thermoregulation, why did these animals have such tall structures, and what does this have to do with Acro?

I suggest that the tall spines in *Spinosaurus* and the high ridges in Acro were for looks: they helped animals to recognize others of the same species, made the animals look bigger and more intimidating in fights for mates and territory, and/or made them look sexy and attractive to a mate. (Beauty was in the eye of the dinosaur.) The reason I make these suggestions is that so much about what we see in the animal world today is directly or indirectly about sex: African antelopes have a variety of horn sizes and shapes, not for protection against lions but to allow them to recognize one another and for males to attract and fight for mates. Male deer grow antlers every year so that males can get into jousting contests to see who will get the females. The male peacock is one of several birds that strut around showing off their tail fans to entice the females to mate. These various body structures have a behavioral purpose, and I suggest the same may have been true of Acro and other tall-spined dinosaurs. This would also explain why the carcharodontosaurid *Concavenator*, known from a nearly complete skeleton, has only two tall spines, which are located just in front of the hips (Ortega, Escaso, and Sanz 2010). Such a small number of spines would be worthless for thermoregulation but would make the animal noticeable.

HOW DID THE TALL BACK SPINES HELP ACRO?

We know that many animals today have territories or home range where they do their thing: eat, sleep, mate, raise young, and so forth. Jim Farlow and Eric Pianka (2002) have estimated that a 2.5-ton (2,270-kg) Acro—which would be a lightweight, immature individual—would have needed a home range of 23,000 square miles (60,000 square kilometers [km^2]; roughly a square 152 × 152 miles, or 244.6 × 244.6 km) if it had a high, mammal-like metabolism but only 7.5 square miles (19 km^2) if it had a low, crocodile-like metabolism that required it to spend most of the day basking. I'll discuss the evidence for metabolism in chapter 6, but regardless of metabolism, animals have to defend their home ranges from intruders, and that means patrolling the turf. If an adult male Acro came across an intruder—say, a young male trying to establish its own territory—the two Acros would probably have tried to intimidate one another before fighting. The danger of fighting is that the attacker is just as likely to be injured as the intruder, and that could be fatal in a world with no veterinarians or antibiotics. Hollywood has it all wrong when two prehistoric monsters immediately launch into attacking each other!

The eyes of Acro are best with side vision because their eyes faced mostly to the sides, so the Acros probably stood sideways to size each other up. It is quite likely that the younger animal would decide that the larger, older

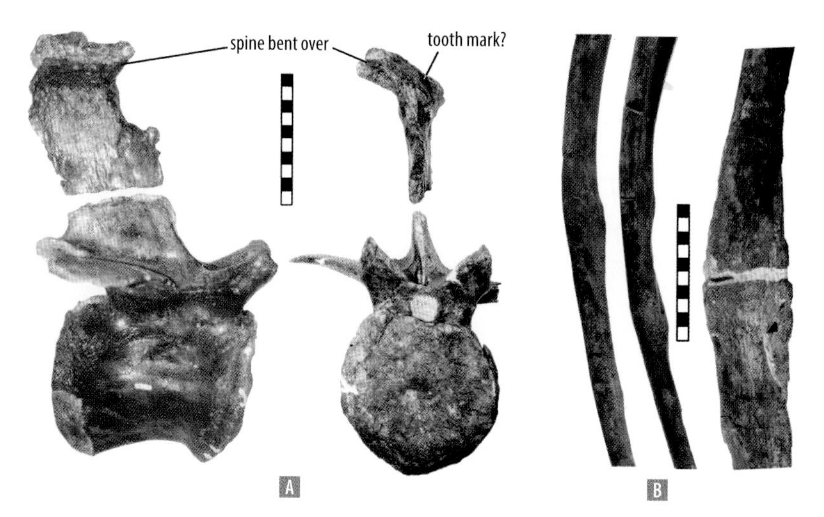

5.4 Bone injuries, or pathologies, can provide clues to behavior. **A** The damaged spine of this tail vertebra (side and front views) of the Texas Acro was interpreted by Jerry Harris (1998) as due to a bite from another Acro. The bite (note possible tooth mark) cracked the spine and folded it over. **B** The same specimen also shows healed broken ribs, although knitting was incomplete for one rib. Scales in centimeters.

animal would probably win in a fight, so the younger animal would leave. In this scenario, having a taller-spined back would be an advantage in making the older animal appear larger and more intimidating. But what if the two animals were more evenly matched in size? In that case, there is the possibility of a fight. Harris (1998) described several bones that had been damaged in the Texas specimen during its life. He reported a depression in a spine that *might* have been caused by a tooth. The bone in this region is also bent over, and regrowth of bone has "frozen" the damaged part at an angle (fig. 5.4A). We would also expect damage on at least one neighboring spine based on the size of Acro's mouth, but unfortunately the tail vertebrae are too incomplete to determine that. The same specimen also shows damaged ribs (fig. 5.4B). Most of them have swollen growths called calluses where the fractures were healed, but one rib has pseudarthrosis ("false joint") where the break did not mend and the two bones remained partially separated.

There is also damage to the shoulder blade of Fran (fig. 5.5). Peter Larson suggested to me (personal communication, 2014) that it may be from a bite, but I have my doubts. It is hard to imagine how a mouth could get to the shoulder blade, considering that this part of the chest had a large surface and

5.5 This shoulder blade of Fran shows damage from a supposed tooth puncture. It is difficult, however, to envision how a mouth could grab hold of the bone because the shoulder blade was embedded in muscle on the side of the body.

the bone was deeply embedded in muscle and skin. In many ways, getting to it would be like trying to bite the center of your palm with your fingers stretched out—you just can't do it. Like the Texas specimen, Fran has broken and healed ribs that might have been the result of a fight, but we can't rule out broken ribs as a result of being whacked by the tail of a large prey animal defending itself. Regardless, injuries do occur, but they do not necessarily tell us what behavior was going on when they occurred.

The tall spines of Acro might also have been used to attract females to mating, announcing, "Hey, baby, check out this glorious set of high spines!" We might expect males to have taller spines than females, but that is not necessarily the case. For example, although antlers do not appear in female deer, horns are present in both male and female bison. The point here is that we have to be careful about sweeping generalizations claiming that the differences between males and females are always apparent (not counting looking under the tail). Furthermore, all of the specimens of Acro are too incomplete to know if one of them (male?) had taller spines than another (female?). It is highly unlikely that the four most complete specimens known are all male (the specimen from Wyoming lacks complete spines), so it seems likely that females had tall spines as well. (Despite Fran's name, we don't know its gender.) Like the male, the female might have used tall spines to make herself look bigger and intimidating. She would need to do this if she was territorial like the male.

CHAPTER 6

Soft, Squishy Stuff

THE INTERNAL ANATOMY

All we have of *Acrocanthosaurus* are its fossilized bones, so it might be surprising that we know anything about its soft, squishy parts that decayed or were eaten. Some soft tissue, such as muscles, nerves, blood vessels, and even the brain, can leave their mark or impressions on bone. For soft tissue that leaves no trace, we can make inferences based on what we know about crocodilians, which are distant dinosaur cousins, and birds, which evolved from earlier theropod dinosaurs. These inferences assume that Acro is evolutionarily positioned, or bracketed, between the two living groups whose anatomy or behavior we understand. Paleontology anatomist Larry Witmer (1997) calls this technique "extant phylogenetic bracketing" ("extant" means living today; it is the opposite of "extinct").

This kind of bracketing has proved to be a groundbreaking tool for making inferences about things that typically don't fossilize, such as soft tissue and even behavior. (Chapter 3 explains the fossilization process.) The technique assumes that if a certain feature is present in both crocodilians and birds, then it most likely also occurred in Acro and all dinosaurs as well. For example, a four-chambered heart occurs in both crocodilians and birds, so it is assumed to have been present in Acro, too, even though we have no direct evidence in the form of a fossilized heart. What phylogenetic bracketing cannot do well, however, is determine whether the heart was more bird-like than crocodile-like, and there are important differences between the two.

Nor can the technique explain characteristics unique to dinosaurs, such as the tall spines of Acro, because neither crocs nor birds have them.

BRAIN AND THE SENSES

Two Acro braincases—the bones at the back of the skull that surround the brain—are known today. One, part of the holotype skeleton, was originally described by Stovall and Langston (1950); it was later described in greater detail by Jonathan Franzosa and Timothy Rowe (2005) from the University of Texas at Austin. The other braincase, belonging to Fran, was described by Drew Eddy and Julia Clarke (2011) of North Carolina State University. Both groups used X-ray computed tomography (CT scans), similar to the scans hospitals use to look inside you. This technique allows the inside of the braincase to be studied without damage. Before CT scans were used, paleontologists had to cut skulls in half and clean out the rock inside. Only then could they get an idea of what the brain structures were like.

CT scans are made by shooting a continuous, narrow beam of X-rays from a source that rotates around the object (person, fossil, etc.). The rotation allows the X-rays to pass through the object at different angles. X-ray detectors on the opposite side gather information about the strength of the X-rays that passed through the object and send that information to a computer. The computer then matches the X-ray signal at each point, or pixel (the same pixels that appear on a computer screen), from different angles to create a two-dimensional slice through the object. The object is moved forward a little after each rotation, and then another slice is made. Each slice is like one frame of a movie, so viewing them in sequence makes it appear as if you're moving through the object from one side to the other. To see some really great examples of this, visit the University of Texas's Digital Morphology website, www.digimorph.org. There, you can see the complete CT scans of the Acro braincase by Franzosa and Rowe.

CT works because X-rays are sensitive to the differences in density, or compactness, of tissue or material. We can understand density by using some Styrofoam packing chips, also known as packing "peanuts" or "popcorn." Take one of these chips and crush it between your fingers as small as you can. This crushed packing chip weighs the same as the uncrushed chip, but it now takes up far less room because you have crushed out most of the air. You could repeatedly crush other chips together until they take up the same space (i.e., have the same volume) as one uncrushed chip. When I did this experiment, twelve crushed chips took up the same space (or volume) as one uncrushed chip. Altogether they weighed twelve times more than the uncrushed chip despite taking up the same amount of room. With X-rays, the denser the material, the fewer X-rays that are able to pass through it because they get blocked by the material. In the uncrushed Styrofoam, there is so much empty space that X-rays would zip right through. But the more empty space we can eliminate, the denser the Styrofoam becomes and the more X-rays that would be stopped by the Styrofoam. This same principle works when a body or fossil is scanned. Soft tissue is less dense than bone, which is why doctors use X-rays to see if you have broken a bone. The X-rays zip more easily through the soft tissue than bone except where there is a fracture. Fossils and their infilling of rock also have variable densities, which is why CT was used on the braincases of Acro. Let's look at what this technique has revealed.

The main purpose of the braincase is to protect the brain. Sealed off from the world, the brain communicates with the rest of the body through nerves that pass through the braincase walls. These nerves are like telephone lines by which the senses communicate to the brain and the brain sends commands to the muscles and organs. Although the brain tissues of Acro are not preserved, the cavity in which the brain sat inside the braincase preserves the rough size and shape of the brain (looking kind of like a bent hot dog, flattened from side to side). By using CT scans, a digital representation of the brain can be created (fig. 6.1). This reconstructed "brain" of Acro has a volume of about 11.5 cubic inches (188.5 cubic centimeters [cc]), which would fit into a cube with 2.25-inch (5.7-cm) sides. For comparison, the spherical human brain's volume is about 88.5 cubic inches (1,450 cc) or that of a box about 4.5 inches (11.4 cm) on each side. Although Acro weighed about fifty times more than the average human, it had only a little more than a tenth of our brain size (fig. 6.2). Put another way, the brain in humans is about 1.5 percent of the weight of the human body, whereas Acro's brain was only 0.004 percent of its body weight! The phrase "dumb as dirt" seems appropriate.

The brain of *Acro* can be divided into two major regions: the cerebrum and the cerebellum. The functions of each region depend on the nerves that feed into them. Let's look at how the brain of Acro would function when attacking and eating prey. At the front of the brain is a pair of long bundles of nerves from the nose. Each of these bundles is an olfactory nerve (biologists call it "nerve I"), and they bring the smell signal from the nostrils to the front part of the cerebrum, called the olfactory lobe, where the signals are interpreted

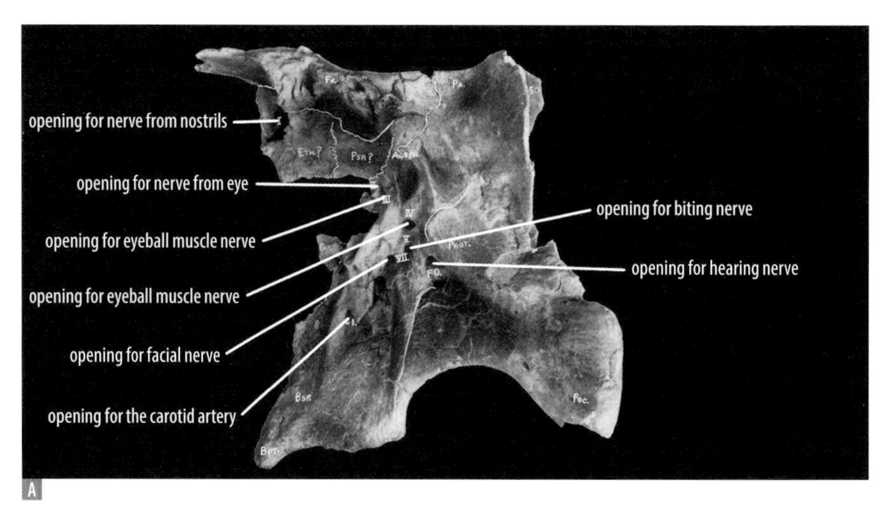

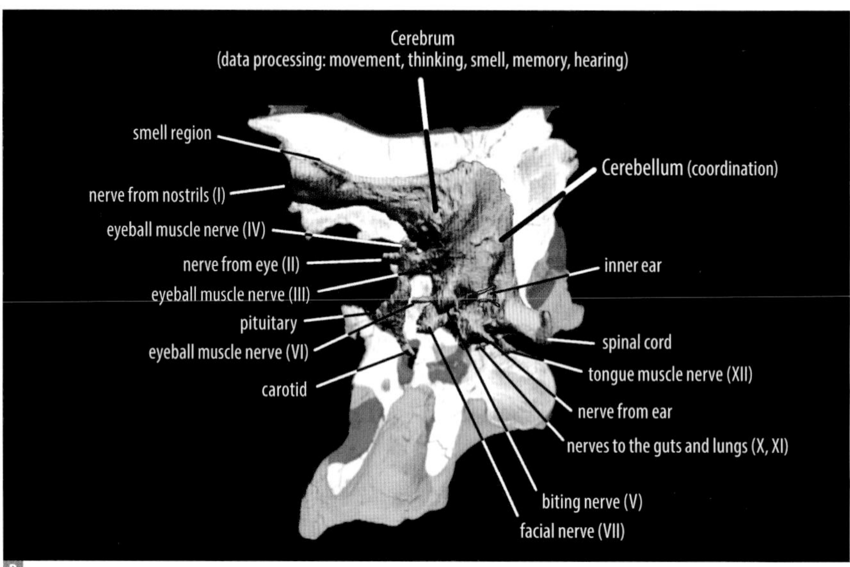

6.1 A Braincase described by Langston (1947), showing the locations of nerve exits. This braincase was CT-scanned by Drs. Richard Ketcham and Matt Colbert at the University of Texas High-Resolution X-ray CT Facility in Austin. B Jonathan Franzosa then used the data to reconstruct the brain cavity, which replicates the shape of the brain and its structures. The results were published by Franzosa and Rowe (2005). *Panel B (in part, excluding the brain overlay) courtesy of Dr. Tim Rowe and Digimorph.org and © copyright Digimorph.org.*

6.2 Relative size of the brains in Acro and in a human. Although the brain of Acro may seem to be almost as large as a human brain, it has much less volume because it is narrow from side to side (see fig. 8.1). The human brain is more of a sphere.

("Smells like lunch!"). The cerebrum signals the legs to start moving in the direction in which the smell is strongest. As the Acro walked, signals to the cerebellum from the inner ear kept the dinosaur from falling over. With each step, the cerebellum coordinates shifting of the body from side to side to keep the weight over the supporting foot (the one that is on the ground).

The nerve from the ear, the auditory nerve, picks up sounds of prey moving and feeding ("Sounds like lunch!"). The cerebrum processes this data and adjusts the direction of travel toward the sounds. Below the olfactory lobe is a small, protruding mass where several different nerves come together at the cerebrum. These include the optic nerve (nerve II) from the eyeball. The cerebrum interprets the signal of what the eyes now see ("Looks like lunch!"). The eyeballs track the prey with signals from nerves III, IV, and VI, which control various eyeball muscles. Meanwhile, a part of the cerebrum compares what the eyes see with recollections of past prey ("Yes, that kind of prey was definitely tasty the last time!"). The cerebrum signals the body to get ready to attack. Once the cerebrum decides the body is close enough, it signals the legs to start running toward the slowest prey animal, which is usually the youngest, oldest, or sickest of the group. Just as the Acro gets to the prey, the mouth opens (signaled through nerve V) and then closes on the fleeing prey. When the mouth closes, feedback from the facial nerve (nerve VII) tells the cerebrum that the attack was successful. The cerebrum then signals the hind legs to stop the forward rush and also signals the arms to grasp the struggling prey. The mouth opens and bites down on the prey's neck, killing it. Lunch may now be enjoyed. The brain signals the guts that food is on its way in order to start the digestive juices flowing and the stomach churning. Flesh cut from the prey by Acro's serrated teeth is pushed around the mouth by the tongue, which gets its instructions from nerve XII to align the food with the throat for swallowing. Once the stomach is full, it signals the brain and the animal stops eating. Now we have a satisfied Acro, which then likely seeks a good spot for a nap.

How good was Acro's vision while it was hunting? Part of that answer comes from the size of the eyeball, which I can estimate from the size of the eye socket. What is less certain is how big the pupil was (the pupil is the black circle in the center of your eye). Animals with good night vision, such as owls, tend to have very large pupils so they can capture as much light as possible. Fortunately, we can use a surrogate to estimate the pupil's size in Acro. Many dinosaurs—perhaps all of them—had rings of paper-thin bones (scleral ossicles) embedded in the iris, the colored part of the eye (fig. 6.3). These rings, called sclerotic rings, also occur today in lizards and birds, where they provide a rigid "shell" within the iris for muscles on the underside to work against. When the muscles are in their relaxed state, the lens may stick out of the pupil, pushing the cornea (the clear, outer part of the eye) forward, and the eye is focused for distance. But when the muscle fibers contract and shorten, several things happen: the inner layer of the eyeball is pulled back toward the pupil, and the lens, which is soft, is squeezed and becomes shorter. All of these actions distort the eye and change the focus to see nearby objects, which for Acro might have been prey. In contrast, humans (and all mammals) have muscles in the eye that deform the lens to change the focus. With aging, the lens hardens so that it does not deform well, which is why older people wear bifocals and even trifocals.

Although we do not have the sclerotic ring for Acro, we can get a fairly good idea of its size by comparing the size of the sclerotic ring in a variety of dinosaurs and other ancient reptiles with the front-to-back length of their eye socket. Using the data given by Schmitz and Montani (2011), Fran, with a 4.5-inch (11.5-cm) long eye socket, is predicted to have a sclerotic ring with an outer diameter of 2.5 inches (6 cm; fig. 6.4). Reptiles as a whole have sclerotic rings that fill about three-quarters of the eyeball. From this I predict that Fran had an eyeball 3 inches (7.6 cm) wide, about the size of a navel orange. In comparison, a human eyeball is about 1 inch (2.5 cm) wide (fig. 6.5), an elephant's is almost 1.5 inches (3.8 cm) wide, and a blue whale's is 5.75 inches (14.6 cm). The biggest eyeball may have belonged to an extinct marine reptile called *Temnodontosaurus* (tem-no-dawn-toe-soar-us; Motani, Rothschild, and Wahl 1999), which had an eyeball estimated to have been a whopping 10.75 inches (27 cm) wide! Thus, the eyeball in Fran was large but certainly not the largest.

Eyeball size in animals depends on several factors that we need to understand in order to put the eyeball size of Fran into context. Obviously a mouse's eyeball is smaller than a housecat's, which is smaller than a person's, which is smaller than an elephant's, but the time of day during which an animal is typically active is also important. Animals that are active primarily in low light, such as at night or in deep water, tend to have large eyeballs to collect what little light there is (Motani, Rothschild, and Wahl 1999). Biologist Lars

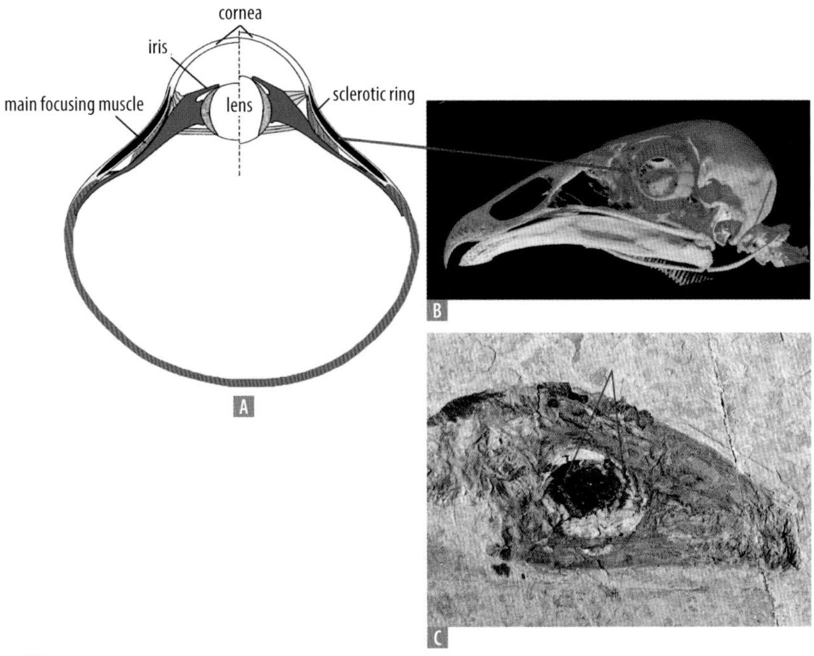

6.3 **A** Section through a bird's eyeball showing the focusing muscles that attach the sclerotic ring to the iris. The right half is the eyeball in the relaxed stage. To focus nearby objects, the focusing muscles use the sclerotic ring as a brace to pull the iris back (left half), which deforms the lens. **B** The sclerotic ring (highlighted here in pink) in a vulture is made from very thin pieces of bone. **C** Sclerotic rings are known in dinosaurs, such as this small theropod, *Sinosauropteryx*. Although no fossil evidence has been found that proves Acro had sclerotic rings, it undoubtedly had them in order to focus. *Panel A based on Walls (1963). Panel B courtesy of Dr. Tim Rowe and Digimorph.org.*

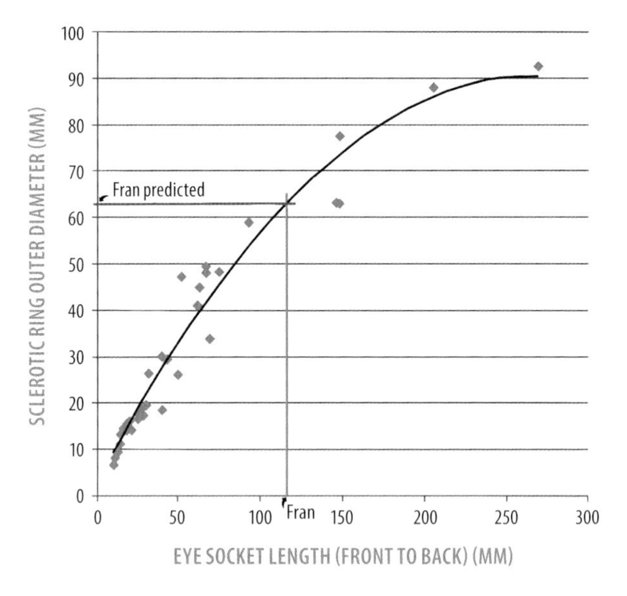

6.4 The sclerotic ring in living animals (blue diamonds) corresponds roughly to their eye socket length. We can use these data to predict the possible size of the sclerotic ring in Fran.
Data from Hall et al. (2011).

6.5 Comparison of a human's eyeballs with the reconstructed eyeballs of Fran.

Schmitz and paleontologist Ryosuke Motani (2011) went so far as to say, based on eyeball size, that predatory dinosaurs such as Acro were active mostly at night. Margaret Hall, an evolutionary biologist, and her colleagues (2011) argued that the story is more complicated and involves more than just eyeball size. I think Hall and her colleagues are right because a separate study recently showed that faster-running animals tend to have larger eyeballs than related but slower-running animals do. That is because faster-running animals need to see and avoid obstacles (Hall et al. 2011). For Acro, the large, 3-inch (7.6-cm) eyeball was partly due to the animal's large size and, as a possible predator, to its need to see and chase prey without running into things or tripping. I suspect that Acro might have had better night vision than humans do because of its larger eyeball, so don't imagine you could sneak around in the dark and be safe from an Acro.

It was also important for Acro, as a hunter, to judge distances to its prey; otherwise it may have closed its mouth on air. Judging distance requires depth perception, which is made possible by stereoscopic or 3-D vision. Depth perception is a neat trick of the brain, which combines the overlapping portions of the fields of view for each eye. (The eyes must be spaced relatively far apart from each other in order for this to work.) To understand this concept, stand up, put your hands down beside you, close one eye, and then bring the tips of your index fingers together about a foot in front of your face—I'll bet you missed. Now repeat, but with both eyes open—you were probably successful this time. This simple trick shows the importance of 3-D vision for judging distance.

In humans, the forward-facing eyes have an area of overlap of about 120°; in dogs it is 60°, and in Acro it was an estimated 40° (fig. 6.6). The wide overlap in humans is a carryover from our tree-dwelling primate ancestors, who needed to judge distances accurately when leaping or swinging from branch to branch. Dogs, Acro, and other carnivores need a narrow zone of vision overlap in front of the snout in order to know exactly when to bite.

This 3-D zone is not the only area of vision, however. In fact, peripheral vision may be more important for monitoring the background. A dog's field of view—all that it can see without moving their heads, including both the overlapping and nonoverlapping parts—is about 240°. In horses it is about 357° because their eyeballs bulge out, giving them more clearance to see if a predator is sneaking up on them from behind (bulging eyes are rare in carnivores). Acro's field of view was a panorama of 280°. In carnivores, a wide field of view enables them to see prey that might have been hidden when the

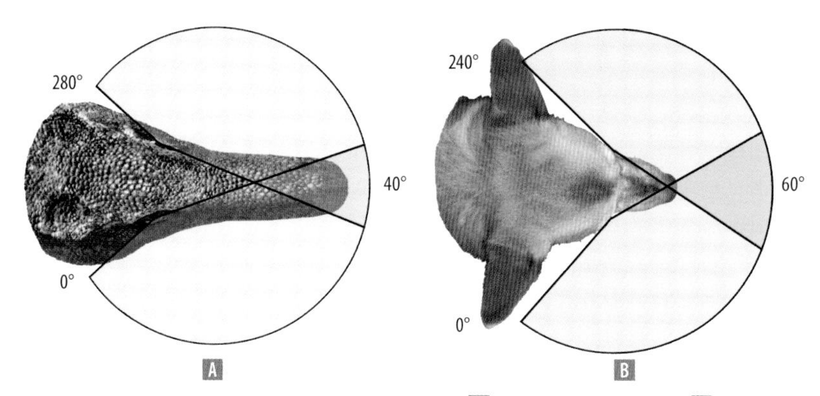

6.6 The field of view—what the eyes can see—in Acro A and in my dog (Tasha) B. The overlapping view at the front makes stereo vision possible, which is important in hunting. Acro and the dog do not need to judge distances to the sides, but they do need to see what is beside or behind them.

predator passed by. Because the world is full of stationary objects—trees, other plants, rocks, and the like—predators tend to tune them out and notice only moving objects. That is why so many prey animals, such as rabbits and fawns, freeze on the principle that if they act like a rock, they won't be noticed. This concept was illustrated in the first *Jurassic Park* movie when paleontologist Alan Grant told young Lex Murphy to hold still at the approach of *T. rex*. The lawyer, Donald Gennaro, was eaten because he moved. Let that be a lesson to you!

HEART

As yet, no one has found the heart of an Acro. One claim (Fisher et al. 2000) of a fossilized heart in a dinosaur was in the specimen of the plant-eating *Thescelosaurus* (thes-cell-oh-soar-us) nicknamed Willo, which is on display at the North Carolina Museum of Natural History along with Fran, the Acro. However, this "fossilized heart" has since been shown (Cleland, Stoskopf, and Schweitzer 2011) to be a concretion in the chest cavity (fig. 6.7). A heart may exist in a juvenile specimen of *Scipionyx* (skip-ee-on-icks), a carnivorous dinosaur from Italy. This individual, variably nicknamed Skippy, Ciro, or Cagnolino, was first described by dinosaur paleontologists Cristiano Dal Sasso and Simone Maganuco (2011). This remarkable specimen preserves a great deal of soft tissue as mineral stains or as mineralized structures.

In the region where the heart and liver would be is an iron-rich patch that may represent the heart. If so, it is too poorly preserved for us to identify structures or to use it as a guide to understand the heart in Acro. We must therefore use extant phylogenetic bracketing to make inferences from the hearts of birds and crocodilians.

Both crocodilians and birds have hearts with four chambers, so it seems most likely that Acro had a similar four-chambered heart. These chambers are cavities within the heart into which blood flows and then is pumped out when the heart contracts, or beats. Flow in and out of the chambers is controlled by one-way valves, which are flaps of tissue inside the heart that are forced closed over the openings by the pressure of the blood. How large was Acro's heart? In crocodiles, which have a low ("cold-blooded") metabolism, the average weight of the heart is about 0.2 percent of the body weight, but in ostriches, which have a much higher metabolism ("warm" or "hot-blooded"), the heart is larger, about 0.9 percent of the body weight (Crile and Quiring 1940). If—and this is a big "if"—the heart of Acro was anywhere

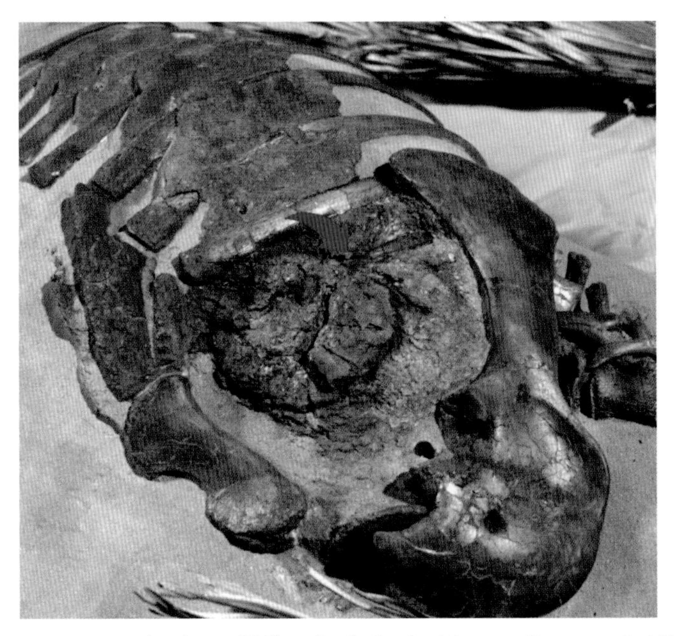

6.7 A large concretion in the chest of Willo, a fossil of an herbivorous dinosaur called *Thescelosaurus*, at the North Carolina Museum of Natural History. The concretion was originally described as a fossilized heart. I cannot rule out the possibility that the decaying heart set up conditions for the formation of the concretion in this part of the chest.

near either of these ratios, then this 5-ton (4,500-kg) animal would have a heart somewhere between 20 and 90 pounds (9 and 41 kg), depending on Acro's metabolism. For comparison, the fist-sized human heart weighs around 0.5 pounds (227 g). I suspect Acro's heart weight was somewhere closer to the middle or high middle of this range because mammals, which have a high metabolism like birds, have hearts that are smaller than those of birds of the same weight. Even at 55 or 60 pounds (25 or 27 kg), that's a big heart! That's as heavy as the heart of a 7-ton (6,350-kg) elephant (Crile and Quiring 1940).

Such a large heart in Acro isn't surprising, considering the results of a study of Fran's bone chemistry that was done by Christine Missell (2004) when she was a graduate student at the University of North Carolina. Missell analyzed the oxygen atoms present in the hydroxyapatite, which is the hard part of bone. Although we tend to think of oxygen as oxygen, there are varieties (called isotopes) that differ in the number of neutrons, which are particles found in the core or nucleus of an atom. An atom of common, everyday oxygen has 8 protons and 8 neutrons at its nucleus, which is surrounded by a moving cloud of electrons. This oxygen, abbreviated by the first letter, is referred to as ^{16}O (the 16 comes from adding the number of protons and neutrons), or "light oxygen." A less common isotope of oxygen has 8 protons and 10 neutrons, which makes it a little heavier, and is referred to as ^{18}O, or "heavy oxygen." Missell found that the ratio of ^{18}O to ^{16}O in Fran was more similar to the ratios found in the bones of a modern elephant and an ostrich than to those in a modern crocodile.

Missell interpreted this similarity to mean that Acro had a high metabolism, more like that of an elephant or ostrich than of a crocodile. In other words, Acro was "warm-blooded" rather than "cold-blooded." Having a high metabolism requires a large heart to pump blood through the lungs fast enough to support the high oxygen needs of the body's cells. These ratio results are somewhat controversial because a variety of conditions can affect the ratios, including the source of the animal's drinking water. (In chapter 9 I'll discuss what oxygen isotopes can tell us about ancient climates and environments.) Still, the conclusion reached by Missell is similar to that of another, more detailed fossil bone isotope study of an Acro relative, *Giganotosaurus*. Geology chemists Reese Barrick (who was Missell's graduate advisor) and William Showers (1999) concluded that the metabolism of this theropod was intermediate between those of mammals and crocodiles but closer to that of a mammal.

LUNGS

A dark smear in the chest cavity of the *Scipionyx* specimen may be traces of its lungs, although the right upper-arm bone hides most of it. Even if that is what the smear is, it is too poorly preserved to provide any clues about what the lungs might have been like in Acro. Other evidence, however, provides clues that the breathing system in *Scipionyx* was more like that of a bird than of a crocodilian. In crocodilians, the liver is used as a piston to push the diaphragm from behind, and this, in turn, inflates and deflates the lungs. (Weird, huh?) Paleobiologist John Ruben and his colleagues (1999) thought that *Scipionyx* had such a breathing system because of the position of the dark smear that was identified as the liver, but this identification was challenged by Dal Sasso and Maganuco (2011), among others.

The lungs of crocodilians and birds differ from those of mammals, which have microscopic dead-end pockets. In mammals the oxygen-rich air is inhaled into the pockets, where oxygen is exchanged for carbon dioxide in the blood; then the oxygen-poor, carbon dioxide–rich air is exhaled. In crocodilians, however, each lung is made up of several compartments or passageways. Inhaled air is pulled into a network of small pouches at the end of the lungs. The liver-piston then pushes the air into the upper passages (called the dorsobronchi) and through several smaller passages to the lower passages (called the ventrobronchi). It is then exhaled. Although air flows in and out of the lung, its movement within the lungs is rather complicated and may be considered, at least partly, to be a one-way movement: the air goes in and out via the same windpipe, the trachea, but within the lungs it makes a one-way loop (Farmer and Sanders 2010; Schachner, Hutchinson, and Farmer 2013). Birds, which have rigid lungs (they don't expand or contract), have a similar but somewhat modified system (fig. 6.8). In addition to the one-way airflow through the lungs, they also have large, thin air bladders, called air sacs, throughout the body cavity. Some of these air bladders also have little pouches, called diverticula, which extend into hollow bones, such as the upper arm bone (humerus) and some vertebrae. Inhaled air is pulled to the rearmost air bladder, which surrounds the guts. From there it is pushed into and through the lungs in one direction to air bladders in the front of the body, and from there is finally exhaled.

What about Acro? It seems a safe bet that its breathing system must have been closer to those of crocodilians and birds than to the two-ways system of

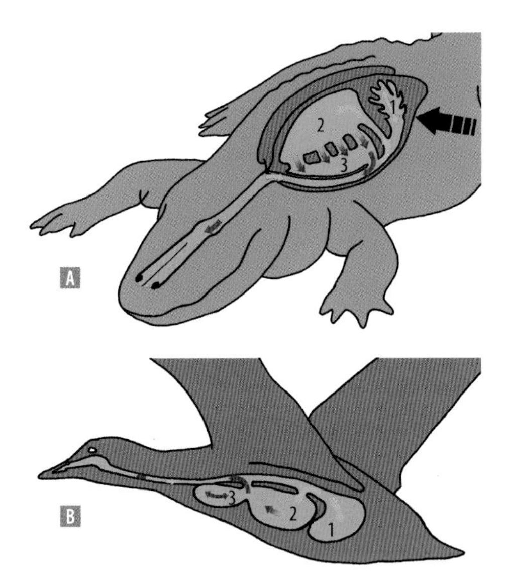

6.8 Crocodiles and birds share some features of one-way breathing. The system in crocs is partially one way and in birds is completely one way. It seems likely that Acro had a breathing system somewhere between those of birds and crocs. **A** In the crocodile, inhaled air (blue arrows) goes to an inflatable pouch in the rear of the lungs. The liver acts as a diaphragm and pushes the pouch (black arrow), forcing the air through the rest of the lungs. There, oxygen and carbon dioxide (red arrows) are exchanged. The carbon dioxide is then exhaled. **B** In birds, inhaled air (blue arrows) is pulled into an inflatable pouch (rear air sac) in the rear part of the body. It is then pulled into the lungs, where oxygen and carbon dioxide (red arrows) are exchanged. The carbon dioxide is then pulled into another inflatable pouch, in the front of the body, and is exhaled from there.

lizards and mammals. There is even some evidence suggesting that the system was more bird-like than crocodile-like because Acro has many openings called pleurocoels (literally, "side hollows") on the sides of many neck and some back vertebrae (fig. 6.9). These openings connect to a network of interconnected cavities. Paleontologists have long thought that these pleurocoels correspond to the pleurocoels in bird bones and may have held pouch-like diverticula from the breathing system (O'Connor 2006). Paleontology anatomists Patrick O'Connor and Leon Claessens (2005) have gone so far as to suggest the body cavities of theropods were lined with bird-like air bladders. In Acro, the openings into the vertebrae extend through all of the neck and back and into the vertebrae of the hips, suggesting that Acro had air bladders to support a lifestyle that demanded lots of oxygen (fig. 6.10). As noted previously, Missell (2004) suggested this lifestyle was that of a highly active predator with a high metabolism.

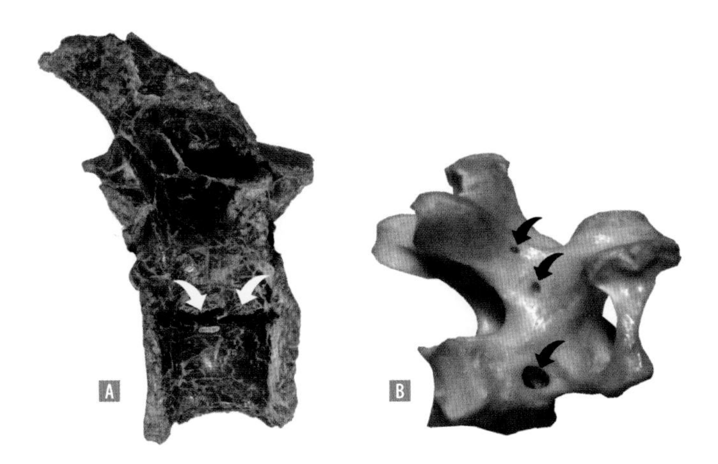

6.9 Openings (arrows) into the vertebrae of Acro **A** and birds **B** suggest that Acro had a more bird-like than croc-like breathing system. In birds, small pouches called diverticula extend from the lungs and air sacs in the bone.

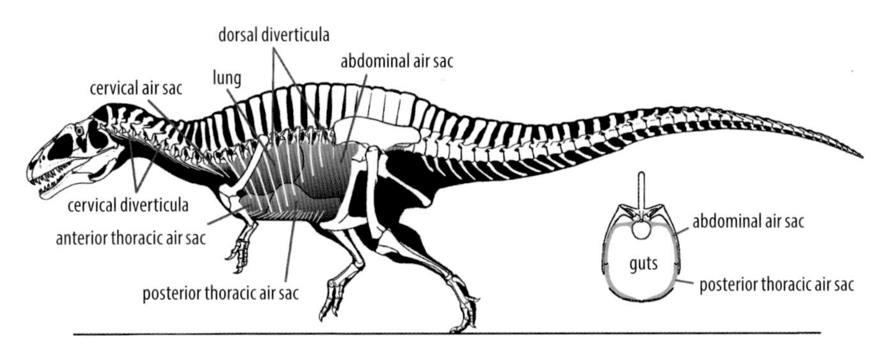

6.10 Possible breathing system in Acro if it had a bird-like breathing system. These air sacs were narrow and occupied the space between the ribs and guts.

GUTS

Although the possible heart and lung are poorly preserved in the juvenile specimen of *Scipionyx* from Italy, some if its organs have been better preserved (Dal Sasso and Maganuco 2011). This remarkable little dinosaur was found in very fine-grained limestone that was deposited in a lagoon, much like the famous specimen of the older *Archaeopteryx* (ark-ee-op-tare-icks) from Solnhofen, Germany. Decaying bacteria deposited minerals in and on many of the internal organs of *Scipionyx*, thereby preserving their shape in fine detail (fig. 6.11). Such unusual fossilization has been shown to occur experimentally on modern organisms by paleobiologist Derek Briggs and his colleagues (summarized in Briggs 2003). The actual tissue decayed but left behind a "ghostly," mineralized shell of many organs. The preserved details are not very good for the heart, lungs, and possible liver, which are all blood-rich organs. It's likely that decay was too rapid in these areas and happened before enough minerals were deposited on them because of the high oxygen levels

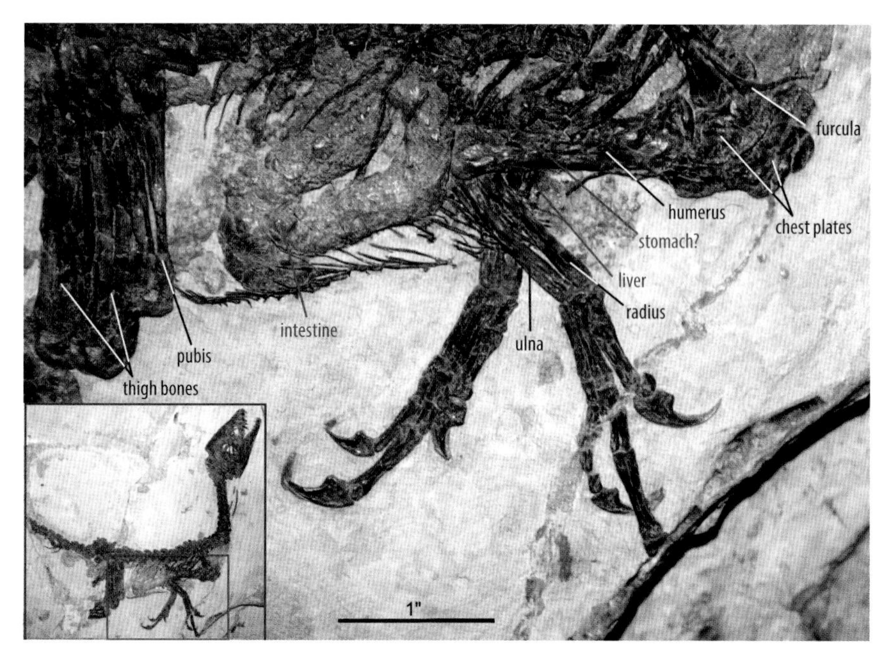

6.11 Fossilized soft tissue found in the small theropod *Scipionyx*. This specimen and similar ones showing soft tissue provide clues to the soft anatomy of Acro. Names in red indicate fossilized soft tissue, whereas names in black indicate bones. *Photo by Giovanni Dall'Orto, courtesy of WikiCommons.*

in those tissues. Oxygen levels are lower in the intestines, where bacteria live that do not need oxygen (called anaerobic bacteria). These "friendly" bacteria in our guts play an important role in digestion—and you thought all bacteria were bad! In *Scipionyx*, the low oxygen levels of the intestines slowed decay way down and gave the bacteria time to deposit minerals. The detail is so fine that the folds on the inside of the intestines are preserved.

The intestines are remarkably short for the size of the animal. Anyone who has cleaned game knows just how long intestines can be. In herbivores, plant material has to sit and digest for a long time in lengthy intestines. The short intestines in *Scipionyx* shows that it was a carnivore because animal protein is easier to digest than plants are, so long intestines are not needed. From this, we can infer that Acro also had rather short intestines because its teeth show that it was a carnivore. We don't know how efficient Acro's digestion was, but phylogenetic bracketing (discussed earlier) suggests that it may have been as efficient as a crocodile's or an eagle's. Both are good at digesting muscle and small bones, but they tend to throw up larger bones. Once, in Wyoming, I found a tail vertebra of a duck-billed dinosaur that showed etching of the surface and rounding of the corners. I suspect it was a bone that was puked up by a *Tyrannosaurus* after having spent a little bit of time dissolving in stomach acids. Such occurrences may be more common than we know, and acid-etched, puked-up bones from Acro may even show up in the Antlers Formation; it is a matter of recognizing them.

MUSCLE

The surface of bone is seldom smooth. Instead, it has rough patches (scars), ridges, and knobs or bumps. These features are more noticeable on older individuals than on young ones because a lifetime of muscles pulling on bone slowly changes the surfaces where the muscles attach (fig. 6.12). The locations of these scars can be matched with those of crocodilians and birds (e.g., Lipkin and Carpenter 2008), and thus the muscles in Acro can be identified. I won't discuss every one of its hundreds of muscles, just the more important ones used in eating, grasping, and walking.

Two important points about muscles need to be understood: First, muscles work by pulling, not by pushing. Pulling occurs as the muscle fibers contract, shorten, and thicken, which is why guys flex their arms to show off their bulging biceps muscles. Muscles on the opposite side of the arm (the triceps) are passive and get stretched; to unbend the elbow and straighten the arm,

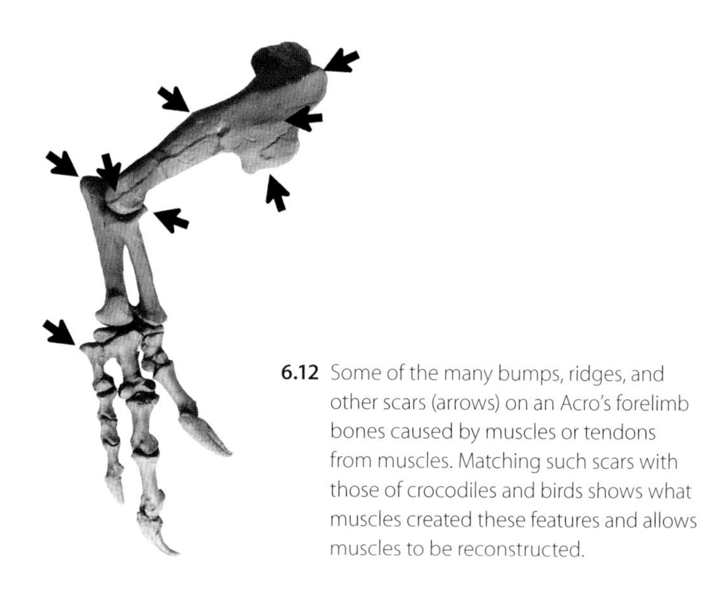

6.12 Some of the many bumps, ridges, and other scars (arrows) on an Acro's forelimb bones caused by muscles or tendons from muscles. Matching such scars with those of crocodiles and birds shows what muscles created these features and allows muscles to be reconstructed.

muscle fibers in the triceps contract and shorten while the fibers in the biceps relax and passively stretch. Second, muscles start on one bone, the anchor, and cross at least one movable joint to attach by a tendon to another bone. (Muscles could not work if they started and ended on the same bone.) When the muscles shorten, they pull the other bone toward the anchor bone. An example of such a tendon can be felt in the crook of your elbow, where the tendon for the biceps muscle crosses to the lower arm. The best way to see for yourself how muscles work is to play with your next chicken before roasting it. You'll need to peel the skin off, then grab a muscle and pull it in the opposite direction from where it attaches. If you pull the front thigh muscle upward toward the hip, the lower leg will straighten. You could do this with cooked chicken, although it won't work as well. If you do it in a restaurant, you'll get weird looks from people at the next table, but just ignore them. They're just jealous.

The jaw muscles, called the adductor muscle group or simply adductor muscles, include several different muscles that work together (fig. 6.13). These muscles are anchored in a box-like cavity called the adductor fossa, which is located between the braincase and outer wall of the skull. You can see into this cavity through the supratemporal fenestra on the top of the skull or through the subtemporal fenestra on its underside (see fig. 4.5). The adductor muscles crossed the joint between the upper and lower jaws to attach to various regions on the inner side of the lower jaw. We can get a rough idea about how powerful the bite was based on the cross-sectional size of the muscles, which can be estimated by measuring the subtemporal fenestra, through which these muscles traveled on the underside of the skull. This opening has an area about 53.5 square inches (345 cm^2) in the skull of Fran. The force created by muscle is about 71 pounds per square inch (psi; 5 kg-force/cm^2). Therefore, the jaw muscles created a whopping 3,800 psi of force (267 kg-force/cm^2) on each side of the head! That is more than enough to bite through the toughest skin and even the smaller, thinner bones of prey. In comparison, a large, adult alligator's jaw muscles combined generate more than 2,100 psi of force (147 kg-force/cm^2; Erickson, Lappin, and Vliet 2003), and human jaw muscles generate a wimpy maximum of 160 psi of force (11 kg-force/cm^2; Scully 2002).

Catching prey required the use of the muscles in the arms and legs. The arm bones of Acro have large attachments for muscles (fig. 6.12); indeed, its shoulder muscles were the largest and most powerful of the arm muscles (fig. 6.14). Three large sheets of muscle—meaning the muscles were wider than they were thick—were anchored along the outer surface of the shoulder blade, crossed the shoulder joint, extended down to the humerus (the upper arm bone), and attached either at a spot on the backside or to the outer surface of a large keel or ridge (called the deltopectoral crest) that projects forward from the humerus. These were the main muscles that pulled the arm back. They were also the largest muscles of the shoulders and arms, so they must

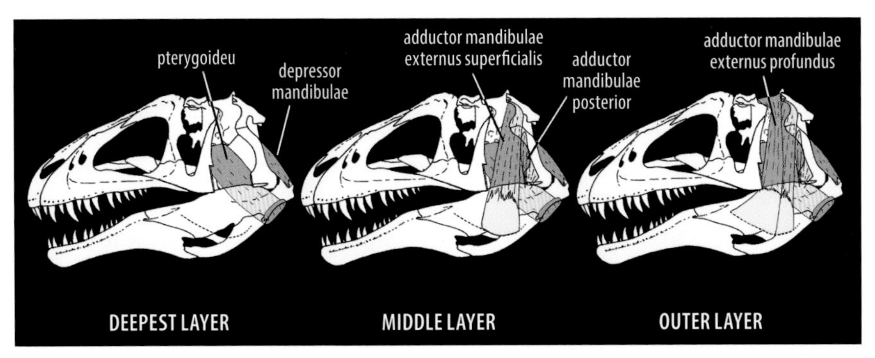

6.13 Jaw muscles of *Acrocanthosaurus* occurred not as a single mass but as several muscles in three layers. The muscle to open its jaw, the depressor mandibulae, did not need to be large because gravity pulled the jaw open. The other muscles, collectively called the adductors, close the jaw.

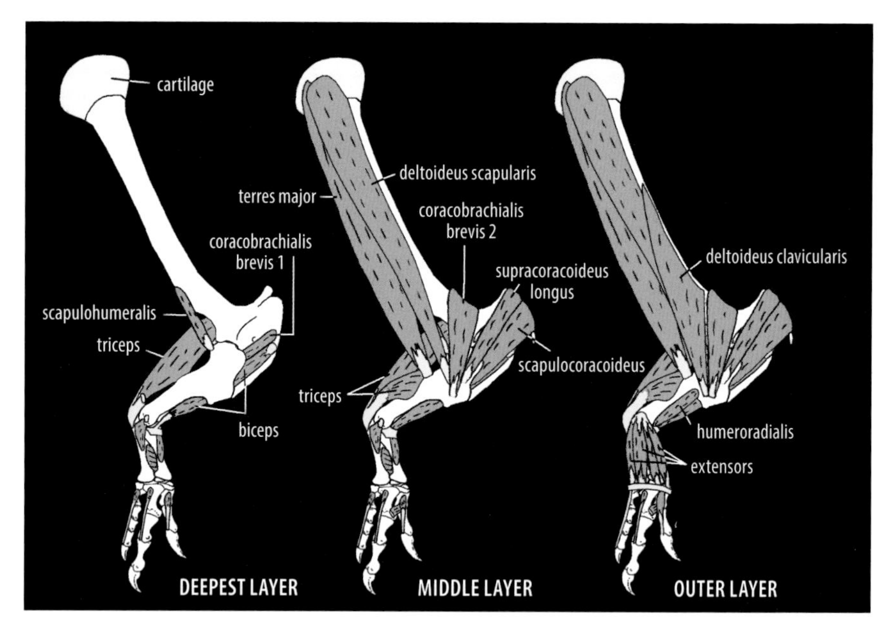

6.14 The arm and shoulder muscles of Acro's right side can be divided into three overlapping muscle groups. These had a complex arrangement: those on the back of the upper arm pulled the arm back, and those at the front pulled the arm forward (to grasp) as well as toward the chest. One important muscle, the pectoral (pectoralis), is mostly hidden by the arm and so was not included in this illustration; it appears in figure 6.15.

have been very important to the animal, undoubtedly to hold struggling prey. Three smaller muscles, together called the triceps, anchored at three different places: two on the shoulder blade and one on the rear, inner side of the humerus. These three muscles joined in one tendon at the elbow and pulled the elbow back, which caused the lower arm to straighten. Another set of muscles anchored on the large, plate-like coracoid bone at the front end the shoulder blade. The coracoid has a large surface for muscles that pulled the arm inward toward the chest and were important for grasping and pulling prey closer (fig. 6.15). The fingers were manipulated by muscles on the lower arm. Those along the outer surface extended the fingers and claws (hence these muscles are called extensors), and those on the inner side (the flexors) closed or flexed the hands and claws when grasping.

As mentioned in chapter 4, the structure of Acro's shoulder joint did not allow the arm to reach forward very far, but it did allow it to move backward quite a bit. It also allowed the arm to move inward toward the chest. These

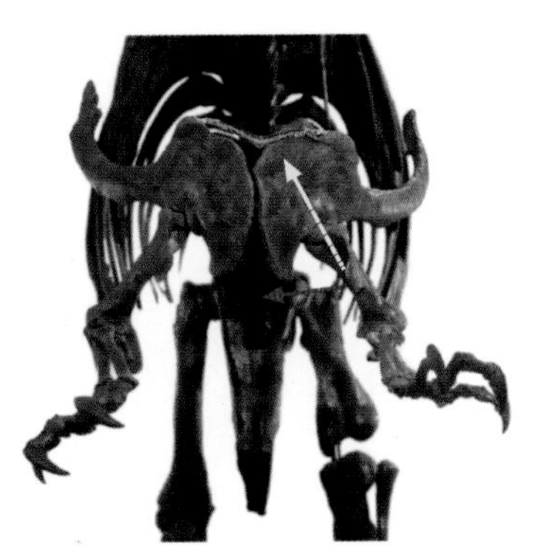

6.15 Front view of Acro showing the narrow chest. Although the wishbone (furcula) of *Acrocanthosaurus* has not yet been identified, it is present in many theropods. The furcula of *Allosaurus* has been digitally added (in pink) to show what it may have looked like. Although small, it would have been an important bone because it binds the left and right shoulder blades together, which is important in holding struggling prey. Two important muscles that pulled the arm inward to the chest during grasping were the scapulocoracoideus (yellow arrow), which pulled upward and inward, and the pectorals (red arrow), which pulled toward the middle of the chest. These two directions of motion are important to pull the prey close to the chest so that the prey can be killed by the teeth.

features are common in dinosaurs and are especially well developed in carnivores, indicating that they did use their arms, even if the arms seem strangely short compared with the sizes of their bodies. In addition to holding prey to bite it, males probably used their arms to hold on to the females while mating.

The hind leg muscles are important for moving the body over the foot that is in contact with the ground (fig. 6.16). In walking or running, muscles on the back of Acro's leg pulled the leg backward, which launched the body forward because of resistance of the foot against the ground. The most important of these was a deep muscle, the caudofemoralis, which was a very large muscle that anchored along the sides of the tail vertebrae and attached to the upper part of the thigh bone. Two other muscles, the flexor tibialis and iliofibularis, anchored on the back portion of the ilium (the upper hip bone) and attached to the lower leg. These two muscles pulled the lower leg back at the same time that the caudofemoralis pulled the thigh bone. But once the caudofemoralis stopped pulling, the iliofibularis and flexor tibialis continued to pull, bending the knee and lifting the foot off the ground. Then several muscles along the front of the thigh bone, such as the iliotrochantericus and iliofemoralis, pulled the hind leg forward. Just as the knee reached its most

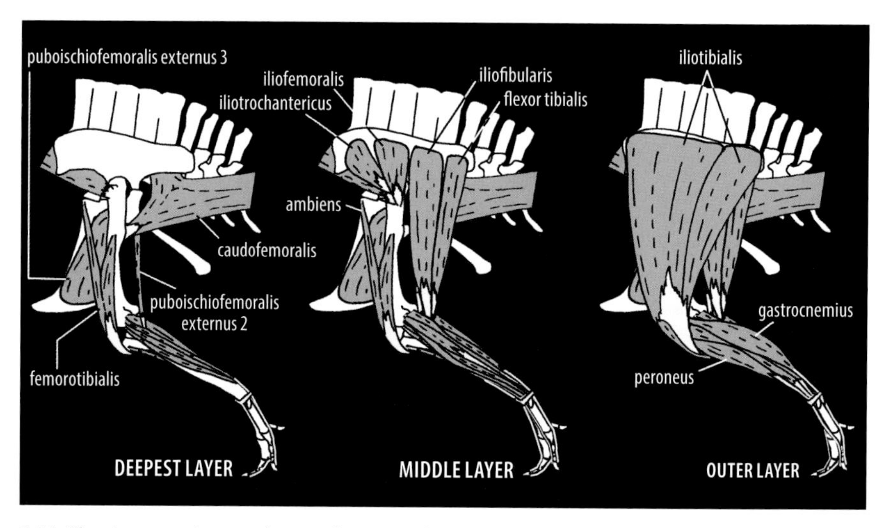

6.16 The three overlapping layers of muscle of the left hind leg and hips. Muscles along the back side pull the leg back and, in doing so, push the body forward. Those of the front side pull the leg forward for the next step. Because it takes more muscle to push the body forward than it does to lift the leg for the next step, the muscles on the back side are larger. (You can see this the next time you eat a chicken thigh.)

forward position, several muscles pulled at the lower leg to straighten it and get the foot into position to touch the ground during the next step.

The toes were operated by several long, slender extensor muscles along the sides of the shin, which straightened the toes, and several flexors on the back of the shin, which were the calf muscles. Another muscle in the calf was the gastrocnemius. This is the muscle you feel in your calf that begins at the back of the knee and is attached to the back of the foot by a long, thick tendon. This muscle was especially important when Acro was running because the body weight on the foot stretched the tendon like a rubber band, and it sprang back when the foot pushed off, giving the body a little more push. To keep the body over the supporting foot (the one that was touching the ground when walking), the puboischiofemoralis muscle along the inside of the thigh pulled the leg inward.

As you can see, even though we have only the bones of Acro to work with, we can infer much about the soft tissue anatomy from birds and crocodilians as well as from the few, rare dinosaur specimens in which traces of the soft tissue have been preserved. I will continue with that line of reasoning in the next chapter, which is about reproduction.

CHAPTER 7

The Next Generation

REPRODUCTION AND GROWING UP

No one has found any eggs or hatchlings of Acro yet, but that doesn't mean we are totally clueless about how the next generation came to be. We can get some insight about what might have happened from studies of modern animal behavior. For example, as discussed in chapter 5, the tall back spines were most likely for show, either to make Acro look bigger to a rival or for a male to impress a female. What if blood vessels under the skin were close to the surface so that the spines could blush? Let's explore why that might have happened by looking at animal behavior during the breeding season.

Imagine that some randy Acro males are cruising, looking for females. If two competitors meet, their immediate response is to size each other up while puffing themselves up to appear more intimidating. We see it with human males in singles bars—guys puff out their chests out and cock their heads a bit to emphasize their chests (sort of like showing off the tall spines), clench their hands into fists, and speak with deeper voices (obviously more manly). Male Acros probably walked in circles around each other to show off their tall-spined profiles, maybe threw in a bit of spine blushing to make sure the other saw how big their spines were, and definitely engaged in a lot of deep growling (or chirping, if they had a bird-like vocal box). Usually, these demonstrations were enough to convince one of the contenders to back down before any major violence occurred. Hollywood would have us believe that dinosaurs always fought to the death, but reality was probably different. Among modern animals, fighting tends to be a last resort because if one animal can inflict injury on the other, the other can do the same. Being seriously injured or disfigured means losing a chance to reproduce.

Sometimes, however, neither would back down, especially if they were evenly matched (in age, size, etc.) or if a younger Acro decided to take on an old bull. Paleontologists Darren Tanke and Phil Currie (1998) noted a high incidence of facial bites made by the sharp teeth of other carnivores in a variety of carnivorous dinosaurs and took that as evidence that carnivorous dinosaurs got involved in some serious roughhousing. The damage on the left side of Fran's skull (fig. 4.5) might be a pathology, an injury, that was the result of a fight because other injuries include some broken ribs are also are partially healed. (The nickname Fran has nothing to do with the animal's gender: Allen Graffham, who owned the specimen at one time, named it after his wife.) Alternatively, the damage could have been done by foolishly attacking a large sauropod; more on this in chapter 8.

Chasing off rivals would have been the first step toward making baby Acros; the second would have been courtship. What would Ms. Acro find attractive? Tall, sexy-looking spines? In the animal world, structures can have dual functions. For example, the dewlap of a lizard can be used to threaten male rivals, but it also serves to entice a female (Carpenter 1999). It is possible that a male Acro danced around a female, showing off his tall-spined back. (Did he also flap his arms like an ostrich?) Showing off spines might be even more reasonable if the scales in this region changed color during the breeding season. We also do not know whether Acros mated for life, as some birds do, or just hooked up during the breeding season. Without a time machine, we're forced to use our knowledge of modern animal behavior in order to speculate about Acro's ancient behavior.

The actual mating act must have been impressive to watch. Most animals mate from the rear, with the female either standing or crouching and supporting at least some of the male's weight. But what would a female Acro do with that long, deep, not-very-flexible tail? It seems that it would get in the way. Perhaps their mating was a cross between crocodile mating and ostrich mating, rather than lion mating. Crocodiles often mate in water, where buoyancy helps, but sometimes they mate on land. There, the male partially climbs onto the back of the female, who is lying on her belly, and positions himself a little to one side to clear her tail. Then he is able to extend his penis from his cloaca, a multipurpose chamber for waste and the reproductive organs, into her cloaca. In ostrich mating, the female squats and the male squats partially on her and slightly to one side. He also extends his penis from his cloaca into hers. From this, I infer that the Acro female also squatted, allowing the male to partially squat on her to get his penis into her cloaca. The business of making an egg comes next.

It is fairly certain that the female Acro laid eggs, although no Acro eggs are yet known. Eggs from several other kinds of dinosaurs have been discovered, some containing embryonic dinosaur bones. Also, birds and crocodilians both lay eggs (extant phylogenetic bracket again), and this means that all dinosaurs probably laid eggs, too. There has been some wild speculation that some dinosaurs might have had live births, but we have no evidence for this yet. Eggs and nests of some other theropods have been discovered, however. Studying them has revealed that they had certain traits in common (Carpenter 1999), and I can use those traits to make an educated guess as to what Acro eggs and nests were like. First, theropod eggs are oval, being longer than they are wide, whereas the eggs of most non-theropod dinosaurs are round, like a grapefruit. Second, the microscopic structure of the theropod eggshell is layered, as it is in bird eggs. In most reptiles and non-theropod dinosaurs, the eggshell is made of tightly packed columns or wedges of calcium carbonate crystals (fig. 7.1). Finally, the eggs were arranged in pairs in a ring large enough for the female to squat among them (fig. 7.2). Remarkably, these features of theropod eggs and egg arrangement also tell us something about female theropods' reproductive organs, which, as soft parts, don't fossilize.

The bird-like structure of the eggshell tells us that the egg-producing system was probably bird-like, not crocodile-like (fig. 7.3). In both crocodilians and birds, the egg starts as a single cell that gets fertilized by the sperm,

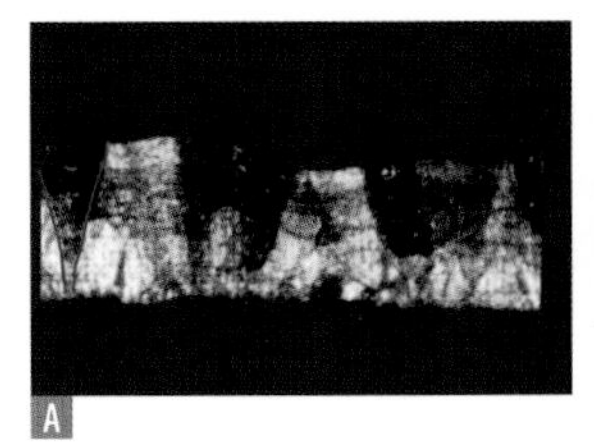

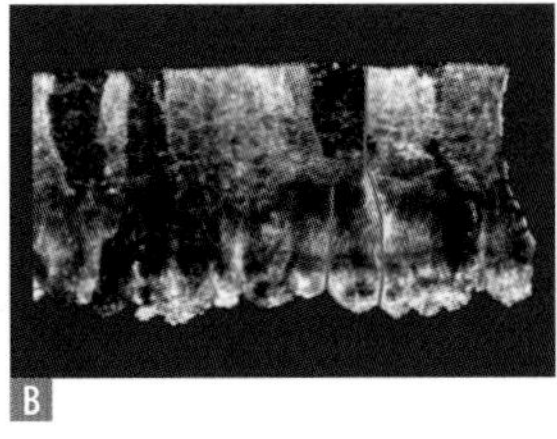

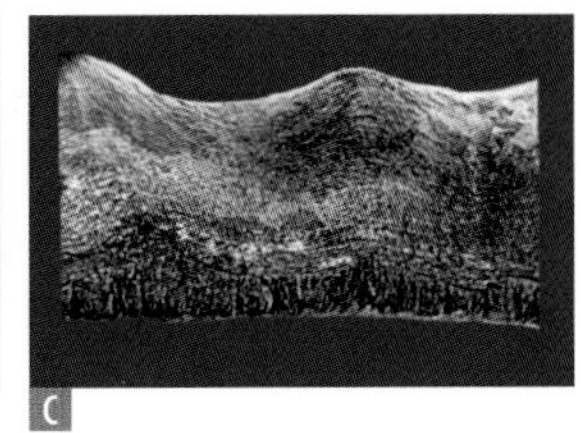

7.1 Eggshells seen through a polarizing microscope, which filters light and, in doing so, reveals the arrangement of the calcium crystals that make up the shell. The outer surfaces of the eggshells are at the top. **A** The crystals in a crocodile eggshell are arranged as wedges. **B** In an eggshell from an herbivorous ornithopod related to *Tenontosaurus* (fig. 8.11), the crystals appear not as wedges but as columns. **C** Theropods have an eggshell that looks like a bird's in that there are no distinct crystals. Instead, the shell is a single or continuous layer. Bumps and ridges often occur in theropod eggshell, as they do in emu eggshell.

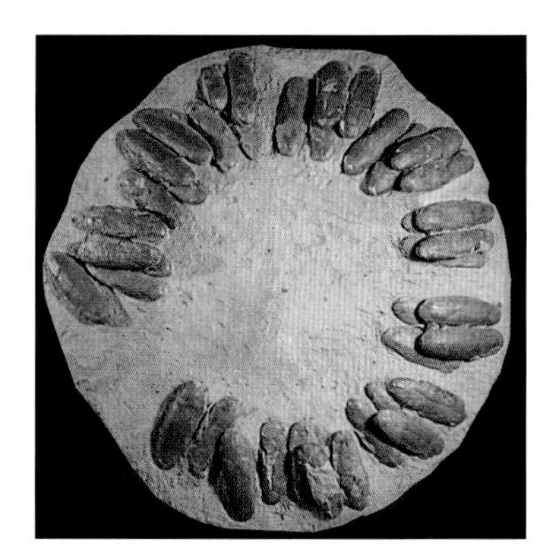

7.2 Eggs of theropods are long, occur in pairs, and are arranged in a ring. Although a clutch of Acro eggs has not been discovered yet, they probably looked much like those of *Gigantoraptor*, pictured here. The open, center area is where the female squatted to lay her eggs. Photo is of a cast of a nest at the Black Hills Institute.

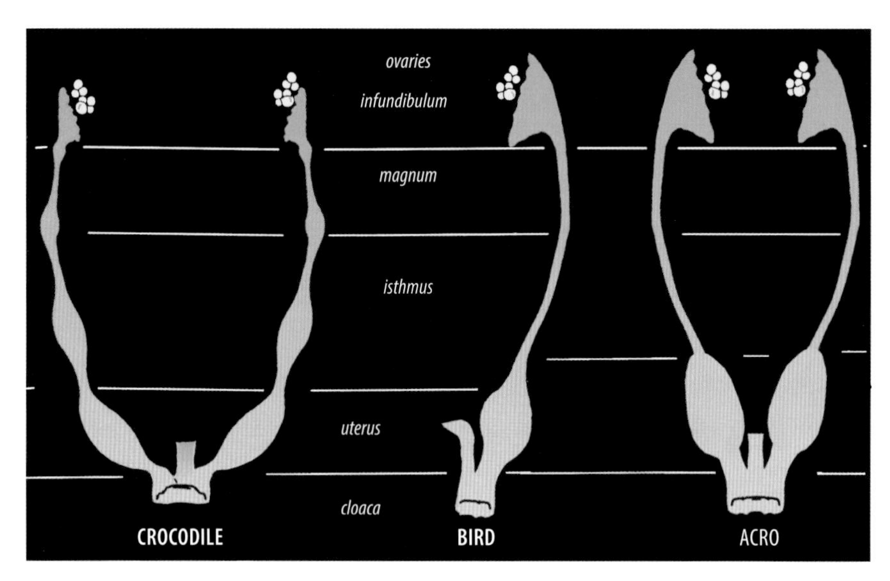

7.3 A comparison of the egg-producing system of a crocodile and bird allows me to infer what the reproductive system of Acro might have been like. The crocodile retains a primitive system of paired oviducts in which the eggs are formed one after another in assembly-line fashion. Birds, on the other hand, have lost one oviduct so as to save weight for flying, and they form and lay one egg at a time. The discovery of paired eggs in the pelvic region of some small theropods in Asia, coupled with the bird-like structure of theropod eggs, suggests Acro had a bird-like, rather than a crocodile-like, egg-producing system. However, Acro probably retained two functioning oviducts. The different regions where various structures are added to the egg are as follows: Ovary—produces the single egg cells, which, once fertilized, will become the embryos; each embryo sits atop a ball of yolk, which is its food source. Infundibulum—the opening into the oviduct through which the ovaries pass. Magnum—the section where numerous glands secrete several layers of albumen (the egg "white"). Isthmus—the part where glands secrete the parchment-like shell membrane. Uterus—shell-secreting glands in this section enclose the egg, thus completing the process.

which had to travel all the way up the oviduct from the cloaca. After fertilization, some remarkable things happen. The egg moves to a funnel-shaped opening, where it begins a long, slow, winding descent through the oviduct. During its travels, various parts of the egg and eggshell are secreted by various glands until, finally, the egg is laid (Carpenter 1999). The first part secreted is a large, yellow blob, the yolk, which is the food for the developing embryo. Next, the egg "white" is secreted in layers around the embryo and yolk. This mostly clear gel protects the embryo and yolk from drying, rapid temperature changes, and jarring. The gel is kept in place by a thin, parchment-like shell membrane; you can see this membrane in a chicken egg and even peel it from the inside of the shell. This shell membrane also provides the surface on which the hard shell is deposited. At this stage, the final shape of the egg is formed by muscle pressure of the oviduct. In birds this happens before the shell layers completely harden, but in crocodilians the eggs (often described as "leathery") never harden completely and are still somewhat flexible when laid.

In crocodilians, the eggs move down the oviduct with a slight time lag so that they are spaced apart. This assembly-line process allows each part of the egg to be deposited before the next egg arrives at the same point in the oviduct. Once the eggs reach the end, they are laid within minutes of each other. As a result, the eggs are laid in a pile. Birds, on the other hand, form and lay only one egg at a time because of the added weight of carrying more than one egg in flight. That means the next egg is not formed until the first egg is laid. For instance, an ostrich will take twelve days to lay twelve eggs. Unlike crocodilians, female birds have only a single oviduct; its partner was lost through the course of evolution, most likely to conserve weight to make flying easier. (Ostriches had a flying ancestor, so they have only a single oviduct.) Remarkably, growth of the embryos in both crocodiles and birds is timed so that all of the eggs hatch at roughly the same time, not days apart. This coordination happens because during the late stage of development, the embryos communicate to each other by chirps and peeps from within their eggs. This apparently signals some embryos to slow the last bit of development while signaling other embryos to speed up development. The result is that eggs laid days apart hatch within minutes of each other.

Carnivorous dinosaur eggs occur in pairs, which means not only that these animals must have had two working oviducts, like crocodilians, but also that each pair must have formed and been laid before the next pair. There is actually fossil evidence for this conclusion. Two different carnivorous dinosaurs—an *Oviraptor* (oh-vee-rap-tore) and *Sinosauropteryx* (sign-oh-soar-op-tear-icks) have been found with paired eggs in the lower parts of their pelvic or abdominal regions (Carpenter 1999). Because there is a rough correlation between a female bird's body size and the time it takes her to make and lay an egg, it might have taken a 5-ton (4,500-kg) female Acro about twelve to fourteen days to lay a clutch of a dozen pairs of eggs (i.e., twenty-four eggs). Once she did, it is doubtful that she sat on them to incubate: at five tons, she would have made omelets! Most likely she buried the eggs under a mound of decaying vegetation, as the crocodile and Australian brush turkey do. The decomposing vegetation would have kept the temperature more or less steady, and it would also keep the humidity high to prevent the eggs from drying out from water loss through pores in the eggshell. To see for yourself how much water an egg can lose, use a food scale to weigh a chicken egg, place it in the sun for a couple days in the summer, and then reweigh it. Pretty surprising, isn't it?

How big were Acro eggs? Size is important because of another correlation: incubation time (the time it takes for the embryo to grow into a hatchling) correlates with egg size. Temperature also matters, so we'll assume an ideal incubation temperature, whatever that might be. Elsewhere (Carpenter 1999), I have given a formula ($12.5x^{0.38}$, where x is body length in millimeters) that correlates hatchling size with adult size and allows us to predict the hatchling size for Acro; that in turn allows us to estimate the egg size. D'Emic, Melstrom, and Eddy (2012) estimated from the growth rings in the bones that Fran had just reached adult size, which is about 39 feet (12 m) long. Because most animals begin to breed before reaching full size (at about 75 percent full size), we'll assume egg-laying in Fran—if Fran was indeed a female—began when she was 29.25 feet long, or 8,915.4 mm. Plugging this length into the formula gives us a hatchling about 15.5 inches long or 396 mm. Folding such an embryo by tucking the head and tail between folded hind legs, like a crocodile or bird, would fit the Acro embryo in an egg 10–12 inches (25.4–30.5 cm) long and 3–4 inches (7.6–10 cm) in diameter, depending on how folded the hatchling was (fig. 7.4). Eggs this size would weigh 1.75–3 pounds (0.8–1.4 kg). That means it took 53–60 days of incubation for the eggs to hatch (for further discussion, see Carpenter 1999, pp. 199–201).

Once the eggs hatched, then what? At this point there is little agreement among paleontologists. Some, notably dinosaur paleontologist John "Jack" Horner (e.g., 2000) and his followers, suggest that there was parental care—that the parents brought food to the hatchlings in the nest because the little

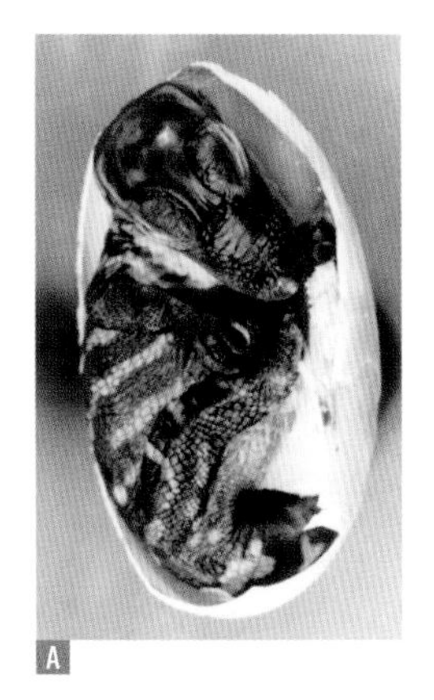

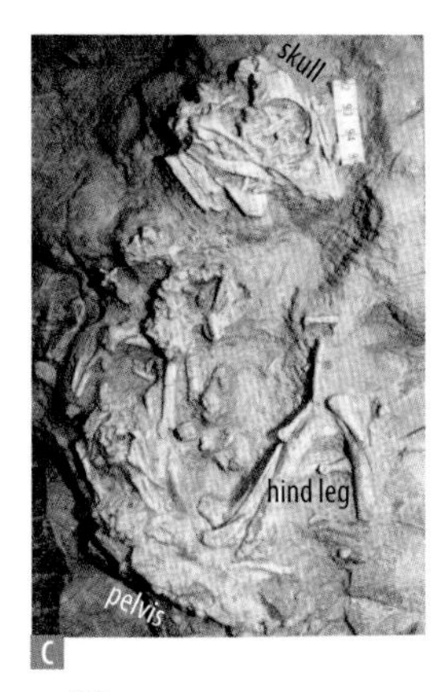

7.4 Embryos just before hatching of A an alligator, B a bird (emu), and C a theropod (*Gigantoraptor*). For the embryo to fit into the egg, its head is tucked against the chest and its limbs folded against the belly. A hatchling Acro probably looked much like these. *Photos A and B by Karl Hirsch, collection of the author. Scale in photo C in centimeters.*

ones were too underdeveloped to hunt for themselves. Such hatchlings, which are called altricial, would line the nest with open mouths, waiting to be fed. Although this presents a cute picture, I have doubts that a parent with a 4-foot (120-cm) head and 3-inch (7.6-cm) teeth could strip flesh from a carcass in pieces small enough to feed these mouths. Other paleontologists, including me, suspect that once they hatched, the little ones would scamper off into the undergrowth, where they would fend for themselves by feeding on insects and small lizards. Such well-developed hatchlings, such as those of ostriches and crocodiles, are called precocial. My argument is supported by very small, juvenile dinosaur skeletons that have been found together but without a parent. In China, the skull of an adult *Psittacosaurus* was supposedly found with thirty hatchling skeletons and offered as proof of parental care (Meng et al. 2004), but this proved to be a fabrication (Zhao et al. 2014).

I suspect that the Acro parents took turns during the eighty- to ninety-day incubation period guarding the nest from various predators, such as large lizards, other dinosaurs, flying pterosaurs, and rat- and cat-sized mammals. Even though the eggs were laid over a span of up to two weeks, the hatchlings emptied the nest within a few hours. Did the parents help by pushing aside the covering over the nest? Possibly. But once out of the nest, the hatchlings scampered to the nearest vegetation for protection from the sun and predators and to feast on insects. Why insects? Because those little crunchy bodies are packed full of easily digested protein and nutrients needed by the little dinosaurs. These hatchlings probably stayed together for the first few months. Given how few young animals today survive into adulthood, the number of Acro hatchlings would have gradually diminished as well, mostly from predators but also from disease. As the remaining little Acros grew, their dietary needs probably changed because it would take increasingly more insects to make a satisfying meal. This problem could be solved by slowly phasing larger animals into the diet, such as lizards and mouse-size mammals, whose fossilized remains have been found in the Antlers Formation (more on this in chapter 9). Eventually, however, the surviving juveniles would inevitably start to view one another as rivals for food and would disperse. From that point on, growing up was a solitary affair.

Until recently, we could only guess how long a hatchling dinosaur needed to grow to breeding age. However, we now have some evidence to back up our speculations. Various bones preserve records of growth, somewhat similar to the way trees form growth rings (fig. 7.5). Called lines of arrested growth, or LAGs, their existence in fossilized bones has been known since the early 1800s. For a while, there was debate (summarized in Reid 1997) as to whether the rings recorded yearly or seasonal growth or were caused by some other metabolic change, such as stress. Since then, however, we've accumulated a large database of evidence, and most paleontologists now agree that many dinosaurs did have seasonal changes in their metabolisms and that these changes produced LAGs. Unfortunately, determining the age of a dinosaur is a little more complicated than simply cutting a bone and counting rings. As a bone grows in length, it also grows in diameter, including the central marrow region. Growth of this inner region destroys the early LAGs, so properly aging a dinosaur requires knowing the number of LAGs that were lost (Reid 1993). Although sometimes paleontologists will make a best guess as to the number of rings that might be missing, a more accurate method is to sample bones from a wide range of different-sized specimens, which are presumably of different ages.

The most relevant scientific study to understanding growth rates in Acro was conducted by paleontologist Paul Bybee and his colleagues (Bybee, Lee, and Lamm 2006) on *Allosaurus*, an Acro relative. They sliced and diced the leg bones of several different-sized individuals from the Cleveland-Lloyd Dinosaur Quarry, located in central Utah. Limiting themselves to a single site allowed them to conclude that only a single species of *Allosaurus* was

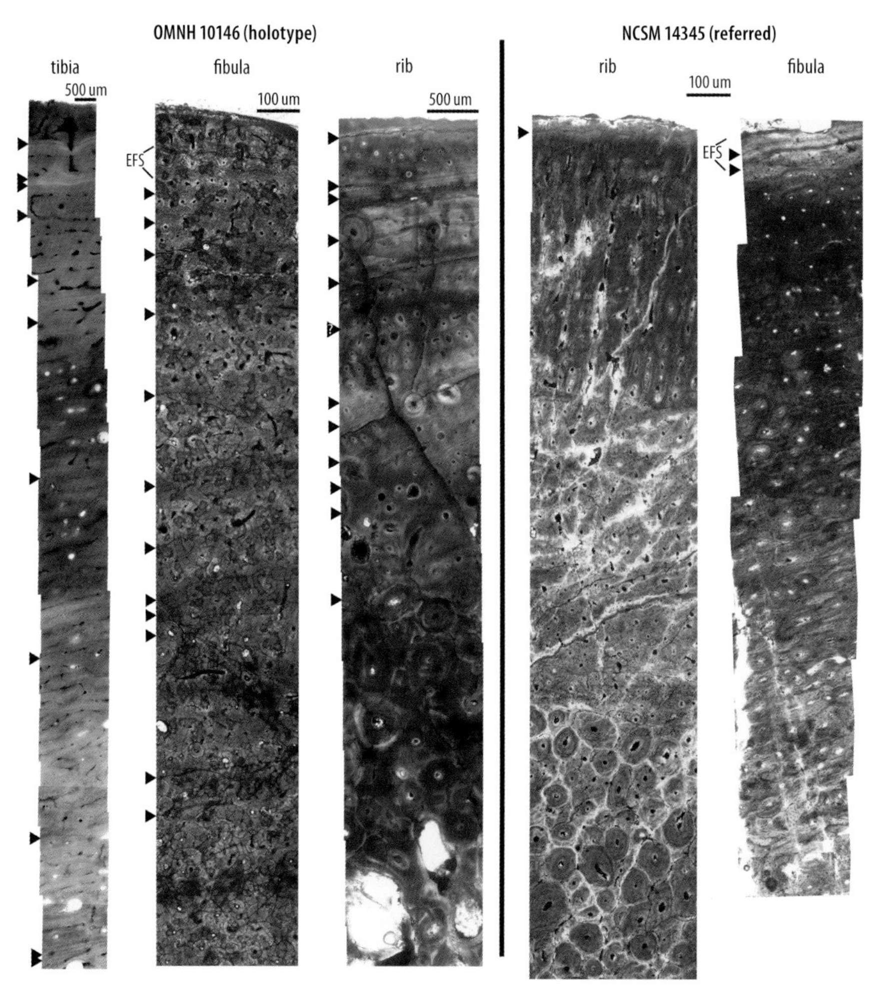

7.5 Microscopically thin pieces of Acro bones provide clues to its age and growth rates from the spacing and numbers of lines of arrested growth (LAGs). Growth was not continuous; it occurred in spurts. Times of slow or halted growth are marked by the lines (shown at the darts). The greater number of LAGs in OMNH 10146 shows that it is much older than NCSM 14345. Note that the LAGs become closer together near the top of OMNH 10146, showing that growth had slowed in the older, more mature animal. OMNH = Sam Noble Oklahoma Museum of Natural History; NCSM = North Carolina Science Museum. *Courtesy of Michael D. D'Emic, Keegan M. Melstrom, and Drew R. Eddy (2012), with permission from Elsevier.*

present, so any difference among the samples was probably not due to mixing proverbial apples and oranges (i.e., mixing different species). By comparing bone length, bone circumference, and the number of LAGs in the arm and upper leg bones, they were able to determine the number of LAGs that were lost as the animals grew. In addition, they were able to assign an estimated age for each individual *Allosaurus*. By plotting the bone circumferences (which increase as the number of LAGs increases) against age, they were able to show that growth slowed with age. They also showed that *Allosaurus* did not live a hundred or more years, as had been assumed when *Allosaurus* was thought to grow like a "cold-blooded" lizard or turtle.

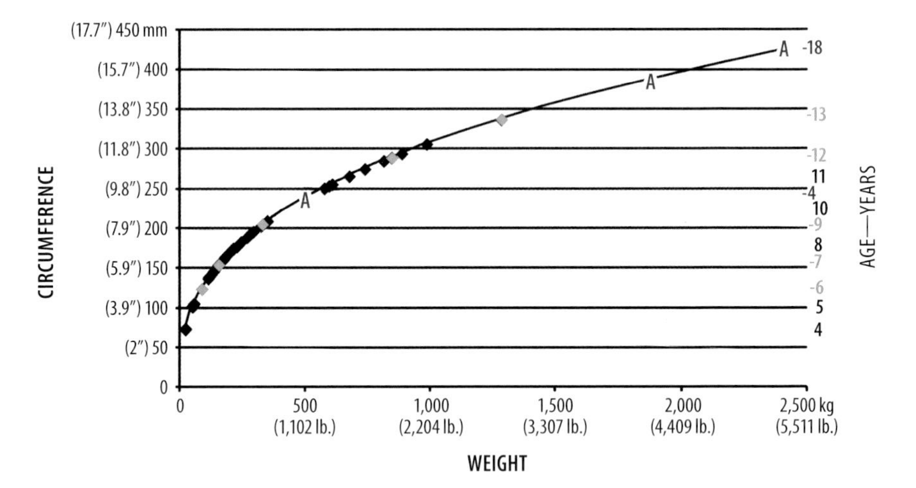

7.6 Bone growth slows with maturity, so instead of counting LAGs, paleontologists can use bone circumference to determine a dinosaur's age. Here, the circumference of the femur is plotted against weight for a population of *Allosaurus* from the Cleveland-Lloyd Dinosaur Quarry. Circumference is also used to determine the probable weight of the individual, and because of this circular reasoning the data points lay neatly along the curve (rather than showing some scatter, as in fig. 6.4). Nevertheless, what is important is that the plot does show the body weight increasing with age, though circumference did not increase as rapidly as it did when the animal was younger. Ages from the LAGs of a sample of *Allosaurus* (Bybee, Lee, and Lamm 2006), shown in green, provide a control for estimating the ages of the various *Allosaurus* and allow extrapolation to the other ages (in black). Plotting Acro data (red), including two ages from D'Emic, Melstrom, and Eddy (2012), shows that it grew more rapidly than *Allosaurus* did. Note the position of a four-year-old Acro compared with that extrapolated for a four-year-old *Allosaurus*. Also, it seems that Acro not only grew faster than *Allosaurus* but also lived longer.

Studies (e.g., Bybee, Lee, and Lamm 2006; D'Emic, Melstrom, and Eddy 2012) have shown that there is a relationship between leg bone circumference and weight. As animals grow and become heavier, their leg bones must get larger as well; otherwise, their bones would break. Using this principle, we can show growth (i.e., aging) by changes in circumference using a large sample of thigh bones from the Cleveland-Lloyd Dinosaur Quarry and plotting circumference against weight (fig. 7.6). Granted, a certain amount of circular reasoning is involved because circumference is also used to estimate weight, but the curve is very similar to that of Bybee, Lee, and Lamm (2006), who used LAGS and fewer specimens. Comparing the LAG age for *Allosaurus* with that for Acro from D'Emic, Melstrom, and Eddy (2012), we can see that Acro grew faster. A six-year-old *Allosaurus* weighed about 240 pounds (109 kg), whereas a four-year-old Acro weighed more than 1,000 pounds (454 kg). This faster rate of growth was comparable to what would be expected if Acro were a giant precocial bird, like a chicken, putting on about 300 pounds (136 kg) per year. Such rapid growth meant that adult size was reached within eighteen to twenty-four years after hatching. How much longer an adult might have lived is not known. Maybe someday a young paleontologist who has read this book will figure that out.

CHAPTER 8

Acro the Hunter

We know from its serrated, dagger-like teeth that Acro was a carnivore, but was it a predator, like an eagle, or a scavenger, like a vulture? Jack Horner and science writer Don Lessem (1994) have argued that everyone's favorite dinosaur, *Tyrannosaurus rex*, was a scavenger, much like a vulture, so is it possible that the *T. rex*–sized Acro was as well? Let's look at the Horner-Lessem arguments and apply them to Acro. In the process we might find out how likely it is that *T. rex* was purely a scavenger.

A CT scan of the skull of *T. rex*, Horner and Lessem argued, shows that its olfactory lobes must have been huge, as they are in vultures. Vultures, as scavengers, depend on finding carrion for food, and they can detect carrion miles away by using the giant smell processors in their brains. Horner and Lessem argued that *T. rex*'s huge smell processors must have allowed it to detect a dead dinosaur far away and, therefore, that this ability points to scavenging lifestyle, or why else would it have evolved? Acro apparently also had large olfactory lobes, based on the impression left on the underside of its skull roof (fig. 8.1). However, unlike the short, wide olfactory lobes of *Tyrannosaurus* (due to it having a short, wide braincase), they are long and narrow in Acro; not surprisingly, the lobes resemble those of its distant cousin *Allosaurus*. As further support that the large olfactory lobes in vultures mean hunting by smell, Horner and Lessem noted that the lobes are much smaller in the eagle, which hunts by vision (fig. 8.2). This is why being able to spot something small at great distance, such as a rabbit hundreds of feet below, is known as being "eagle eyed."

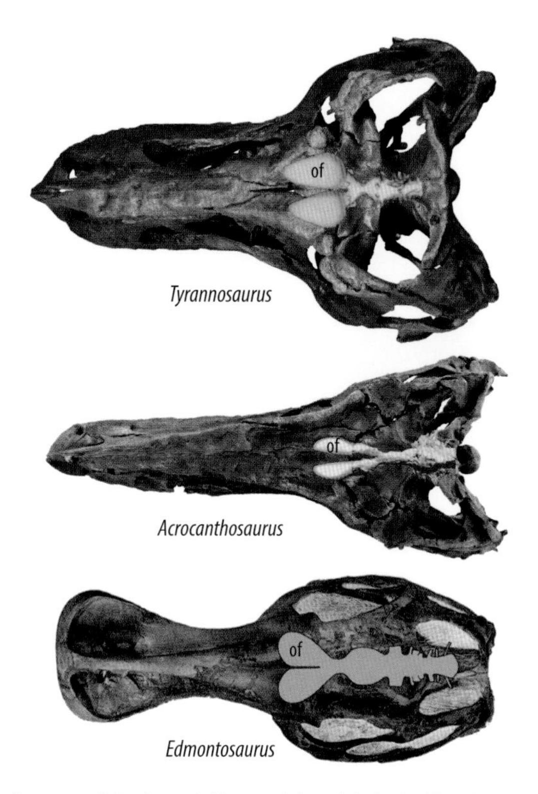

8.1 The large smell region of the brain (olfactory lobes, labeled "of") in *Tyrannosaurus* (*top*) was used to suggest that it was a scavenger. On that basis, Acro (*middle*) must have been one, too. However, the large olfactory lobes in the herbivore *Edmontosaurus* suggest that sense of smell alone cannot determine dietary behavior.

On the surface, there does seem to be a connection between the size of the olfactory lobe and scavenging, so we would predict that large lobes occur only in scavengers. We can test this hypothesis, or prediction, by looking at the sizes of the olfactory lobes in a variety of animals whose diet is known. What we find is that crocodiles have large lobes, as does the albatross, a large seabird that feeds on fish, squid, and penguins (Cherel, Weimerskirch, and Trouvé 2000; fig. 8.2). It would therefore seem that olfactory lobe size alone does not dictate a scavenging diet for living animals or for *Tyrannosaurus* or *Acrocanthosaurus*. What really nails this in dinosaurs is that large olfactory lobes are common among a variety of dinosaurs, not just carnivores, and may be characteristic of the group as a whole (Carpenter 2013). For predatory

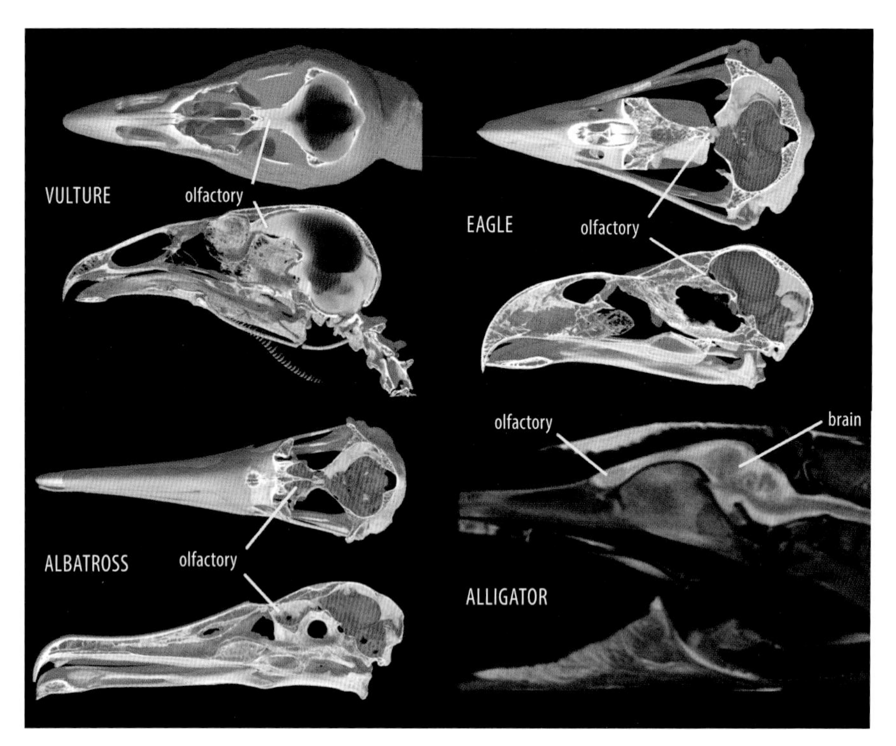

8.2 The smell region (olfactory lobes) in a known scavenger, the vulture, is larger than that in the eagle, which is a visual hunter. However, it is very large in the albatross, a hunter, as well as in the alligator, a hunter and part-time scavenger. The lesson is that it is important to look at a variety of animals when developing a hypothesis (tentative explanation) for extinct creatures. *Bird images from CT scans courtesy of Tim Rowe and Digimorph.org.*

dinosaurs, a good sense of smell may have evolved to locate a mate as much as it did to locate live prey. For their potential prey animals, large olfactory lobes might have been for smelling the bad breath of an approaching predator.

Horner and Lessem (1994) also stated that *Tyrannosaurus* had "beady, little eyes" (p. 161). While it is true that its eye socket is small compared with its skull, which is also the case for Acro, the size of the eye socket in both dinosaurs suggests an eyeball the size of a navel orange (see chapter 6 for how eyeball size is determined). That is hardly a "beady, little eye"; most likely both predators had fairly good vision for tracking prey.

Finally, Horner and Lessem claimed that because *T. rex*'s shin bone is shorter than its thigh bone, it was a relatively slow runner and unable to outrun its prey. Therefore, *T. rex* must have scavenged because carrion cannot get

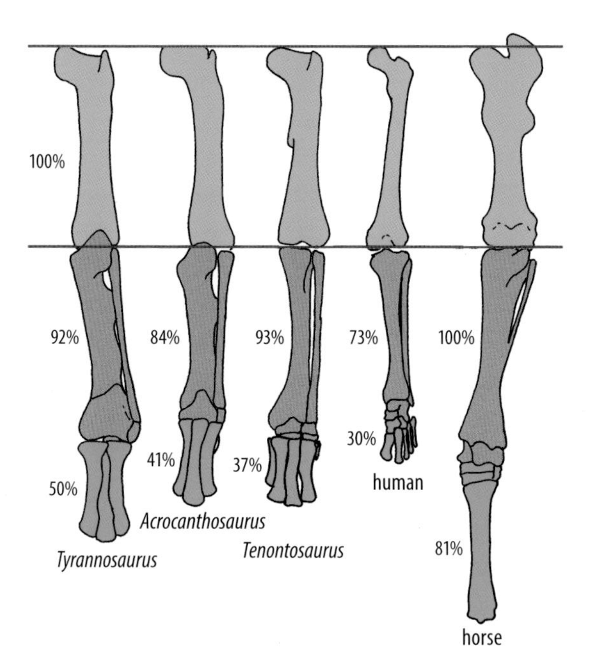

8.3 Comparison of the left hind leg shows that *Tyrannosaurus* and a variety of other species, including humans, have lower legs (pink) that are only a percentage of the length of the upper leg (femur, blue). In the fast-running horse, the two parts are the same length. This difference was used as evidence to support the idea that because humans are slower than horses, *Tyrannosaurus* must have been slow, also. However, horses and *Tyrannosaurus* (but not humans) run on their toes, so the length of the upper foot (green) must also be taken into consideration. Perhaps *Tyrannosaurus* was not slow after all. The proportional length of Acro and its potential prey *Tenontosaurus* are also shown.

up and run away. We know that Acro's shin was also shorter than its thigh, so, according to their argument, it too must have been a scavenger (fig. 8.3). What is the significance of the shin-to-thigh ratio? Among running mammals today, such as horses, and birds, such as ostriches, the shin is as long as the thigh. In contrast, the ponderous elephant has a shorter shin compared with its thigh. Why is this correlation between shin bone length and thigh important? A long thigh bone moves at a slower speed than a shorter one does because there is more of it and the end (knee) travels a longer distance. To see that this is so, take a two-foot piece of string or heavy thread. Tie or tape a weight on the end (washer, coin, fishing weight, etc.). Hold the opposite end and get the weight swinging back and forth like a pendulum. Now hold the string halfway and repeat the swinging. You'll note that the motion is faster. This is the same principle in the hind legs of animals: the knee of the proportionally shorter thigh of horses does not travel as far compared with that of the elephant. Therefore, the reasoning goes, Acro and *T. rex* must have been slow like the elephant, rather than fleet-footed like a horse. But there is more to the story than simply the ratio between two bones. Acro and *T. rex* walked and ran on their toes like birds, so the foot length needs to be taken into account as well. When we do so, we find that the upper foot bones, the metatarsals, are much longer than they are in many (but not all) plant-eating or potential prey dinosaurs.

The amount of ground covered with each step may be far more important than mere shin bone, thigh bone, foot bone ratios. For example, in figure 8.3, the combined lower leg and foot ratio of Acro is 125 percent more than the upper leg, and that of its prey, *Tenontosaurus*, is 130 percent. However, an adult Acro has a leg (hip to toe) that is around 8.25 feet (2.5 m), whereas an adult *Tenontosaurus* has a leg length of only 4.5 feet (1.3 m). With its longer legs, Acro could outrun *Tenontosaurus* because it could cover more ground with each step (fig. 8.4).

How fast could Acro run, or *Tyrannosaurus*, for that matter? That is a tough question for which there is no definite answer. Ideally, we would measure the time it takes for Acro to cross a known distance, such as from one goalpost to another. But given the impossibility of that, we must use indirect means to draw conclusions. For example, widely spaced footprints might have been made by either a running animal or an animal with really long legs. Dinosaur footprint specialist James Farlow (1981) calculated the speed of a possible young *Acrocanthosaurus* running from footprints (about 11 inches long, or 28 cm) in Texas. The set of five prints suggest the animal was running at 26.5 miles per hour (42.6 km/hr). What Farlow admits not knowing is whether the animal was going at top speed or just jogging along. A computer study by biologist Karl Bates (2009) and his colleagues (Bates et al., 2010), in which they reconstructed the muscles and joints of an adult *Acrocanthosaurus*, suggested maximum running speeds between 8.75 and 16.75 miles per hour (14 and 27 km/hr). This is far slower than the animal that made the Texas tracks; the discrepancy is due to Bates et al. basing their model on a larger, older individual. The broad range in speed by Bates and his colleagues shows the uncertainty of current computer modeling because there are just too many unknowns. As the saying goes, “garbage in, garbage out,” meaning that the results of a computer analysis are only as accurate as the information that is put into it. For comparison, estimates on *Tyrannosaurus* speeds range

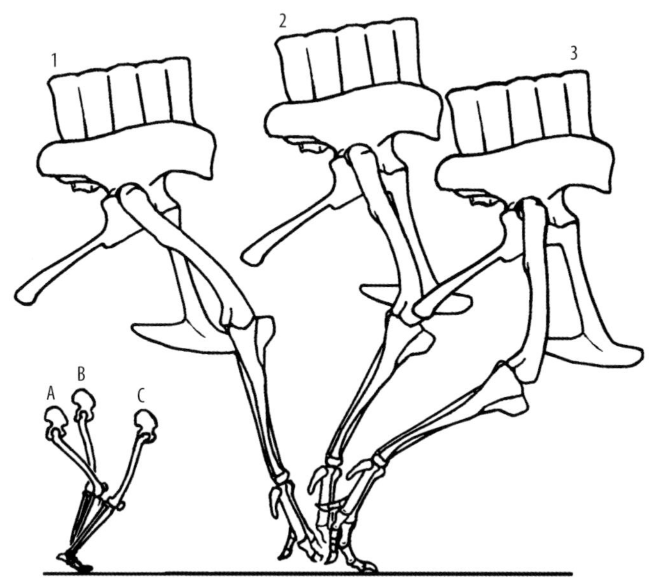

8.4 Comparing the distance the hips can move (right side of the body) in a single step of Acro and a human from the moment the foot touches the ground (1 and A) until the foot begins to be pulled up with the step (3 and C). At midstep (2 and B), the hip is at its highest above the ground. Because the hip (and hence the body) of Acro moved farther with each step, it was essentially faster than a human. However, because the human is smaller, it is more nimble and could have dodged an Acro. So, if you find yourself being chased by an Acro, keep a large tree between you and the beast. The left legs are not shown because the picture would be too cluttered.

from 11 to 31 miles per hour (18 to 50 km/hr). Regardless of the top speed of a running Acro (or *Tyrannosaurus*), all that really counts for a predator is being able to run faster than its prey. As long as it could cover more ground with each step, no prey stood a chance unless it was really huge and too difficult to kill!

Another feature of Acro's feet must have affected its running speed, but it wasn't included in the Bates et al. (2010) model: the energy storage in the Achilles tendon along the back of the foot. As weight was added to the foot, it flexed at the ankle, which stretched the tendon like a rubber band along the back of the foot. Then, when the Acro stepped forward, this energy was released, pushing the animal forward (fig. 8.5). You can think of this as having a pogo stick strapped to each foot. We also must remember that there was a long tail stretched out behind the body. That tail would bounce up and

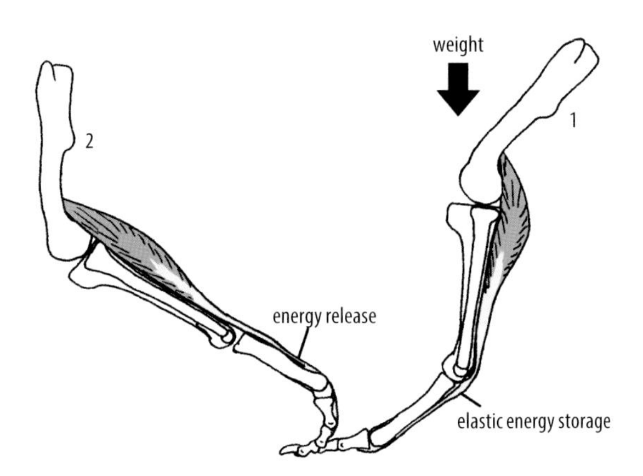

8.5 The Achilles tendon along the back of the leg provided energy storage when the weight of Acro stretched its tendon (1). This energy was released when the weight was removed and helped to push the animal forward (2). Only the left side of the animal is illustrated here.

down with each step as well. As it did so, it would store and then release that energy as it bounced, and that would add even more forward speed. Because the tail was also swishing from side to side as the pelvis moved, the tip of the tail would actually move in a horizontal figure eight.

Acro's walking style is revealed by fossilized footprints, which are fairly common at Dinosaur Valley State Park near Glen Rose, Texas (fig. 8.6). Superficially, these prints look like bird tracks: the three long, slender, spreading toes end in pointed tips made by the sharp claws of the feet (Farlow 2001). Most of the tracks show animals walking across a limey mudflat, and one set has been interpreted as indicating an Acro stalking a large, long-necked dinosaur called a sauropod (a really huge animal, far larger even than an adult Acro; Farlow 1987). This set of tracks, called a trackway, parallels the trackway of a large sauropod (fig. 8.7). On a map of the site (fig. 8.8), one left Acro track is missing based on the pacing of the tracks. Why it is missing is controversial. James Farlow, who has long studied the Paluxy River tracks, suggested that the track-making Acro had leaped onto the left side of the sauropod it was stalking or that it was momentarily lifted off the ground when it struck at the side of the moving sauropod (Farlow 1987; Thomas and Farlow 1997). Personally, I doubt both scenarios because, except for the one missing track, there is no change in the pacing of the steps. In other words, the distance between the tracks is what would be expected of a walking Acro.

To see for yourself why this matters, walk across a room or down a hallway and, at some point, leap into the air and then continue to walk. I'll bet your two feet landed close together to catch and support your weight and to

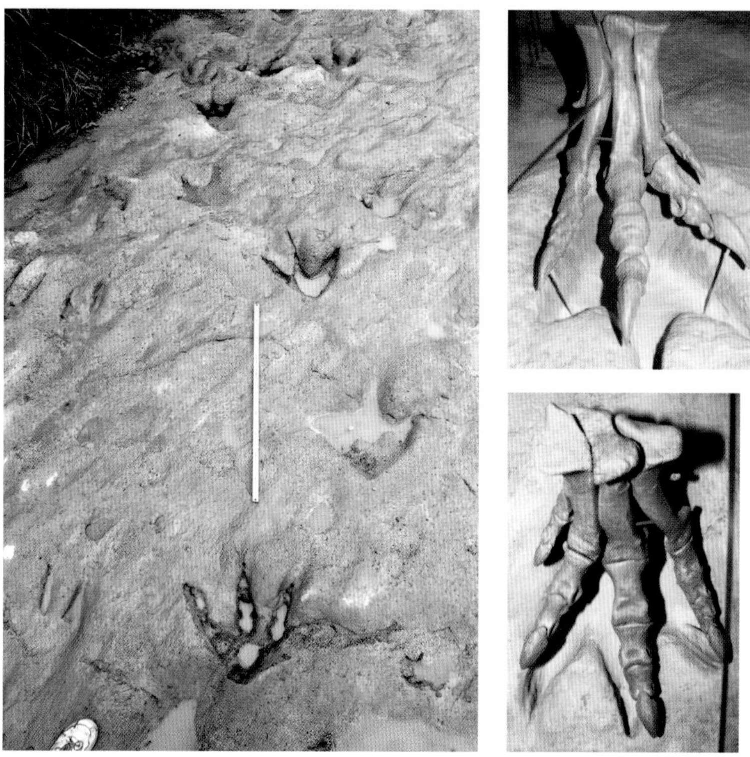

8.6 *Left:* Tracks of *Acrocanthosaurus* along the Paluxy River at Dinosaur Valley State Park in Texas. A meter stick is shown as scale. *Right:* Cast of an Acro's foot bones in a track show how well they match, thus confirming the tracks belong to Acro. *Tracks image courtesy of James Farlow.*

maintain your balance. Imagine how much more important that would be for a 5-ton (4,500-kg) Acro. It is unfortunate that we do not have the end of the story. These footprints continue into the riverbank, where they have not yet been uncovered. What is cool about the tracks along the Paluxy River is that they show exactly where Acro walked. Standing in the footprint of an Acro is to stand exactly where the dinosaur stepped. Unlike bones, which can be moved away by water, a footprint cannot because moving it would destroy it.

The stalked sauropod was actually one of at least four individuals, and possibly more, walking in the same direction. We presume that the animals were walking side by side because the tracks are mostly parallel to one another. We can therefore assume that this was a small herd of sauropods traveling together. This raises an interesting question regarding the supposedly

8.7 Section of the "stalking" Acro and sauropod tracks on display at the American Museum of Natural History. The tracks are deep, indicating the animals walked on soft mud.

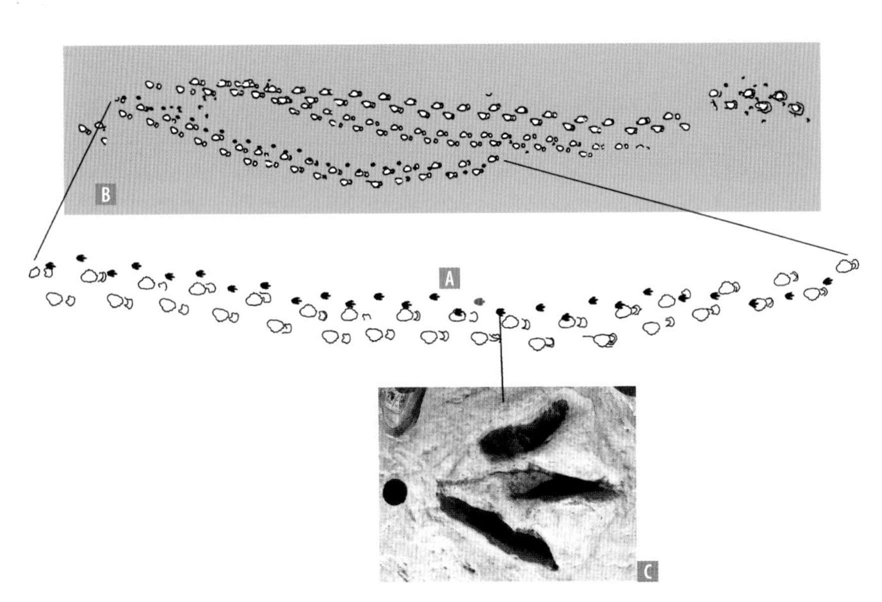

8.8 Map showing **A** the stalking Acro and sauropod in context with **B** other sauropod tracks. The approximate location of the expected track of the Acro (the missing track) is shown in red. This missing track led to the suggestion that the Acro leaped on the sauropod. However, the spacing of the other tracks does not show a change in stride, which would be expected if the animal leaped. **C** An example of one of the Acro tracks. *Map A redrawn and modified after Farlow (1987).*

stalking Acro: why did it not go after a smaller, more vulnerable individual? We can see that one set of prints is smaller, so they were presumably made by a smaller animal. Without a time machine, we will probably never know what really happened or how it all ended.

If these tracks really are evidence of Acro, the predator, stalking a sauropod as prey, who is this other dinosaur? We know that at least some of the sauropod tracks in the Paluxy River valley were made by *Cedarosaurus* (cedar-oh-saur-us) because Mike D'Emic (2012) has identified a hind leg of that dinosaur that was found in the same deposits as the tracks. Other tracks may have been made by the gigantic, really long-necked *Sauroposeidon* (saur-oh-poh-sigh-don), *Paluxysaurus* (pal-ucks-ee-soar-us), or the recently named *Astrophocaudia* (astro-fo-caud-ee-ah) from north-central Texas.

These three sauropods have huge, elephantine bodies, with long necks and tails (fig. 8.9). Their peg- or spoon-like teeth indicate diets of plants, although paleontologists have not yet figured out which plants they ate. *Sauroposeidon* is the most unusual of the three in that it has very long neck vertebrae; the longest, near the middle of the neck, measures 4.75 feet (1.4 m) long (Wedel, Cifelli, and Sanders 2000). As with *Acrocanthosaurus*, its neck and back vertebrae have pleurocoels (openings) in the sides of the vertebrae. Also like *Acrocanthosaurus*, in *Sauroposeidon* these pleurocoels connect to a

8.9 Reconstructed skeleton of the long-necked sauropod *Paluxysaurus*. This skeleton on display at the Fort Worth Museum of Natural History shows the long neck and tail that characterizes the group. The skeleton is 12 feet (3.7 m) at shoulders and is 60 feet (18.25 m) long.

8.10 *Paluxysaurus* being attacked by a foolish young Acro. The scene is set in the coastal environment near what is now Dinosaur Valley State Park near Glen Rose, Texas. Small pterosaurs chase insects.

network of chambers inside the vertebrae that likely were connected by air sacs to the breathing system. It seems rather odd that such huge creatures may have breathed like birds, but it may have been the only way to get air down such long necks (Wedel 2003).

An adult *Sauroposeidon* had a 39-foot (12-m) neck, which is as long as an entire adult Acro! Matt Wedel and his colleagues (Wedel, Cifelli, and Sanders 2000), who first described *Sauroposeidon* from some neck vertebrae collected near the first Acro discovery, estimated that an adult might have been 92 feet long (28 m), stood 23 feet (7 m) at the shoulders, and weighed 55 tons (49 tonnes). It is difficult to envision an Acro trying to take down such a large animal or any adult sauropod, for that matter (fig. 8.10). On the other hand, like every other dinosaur, *Sauroposeidon* started very small when it hatched from an egg, and it would have been vulnerable to a hungry Acro for a good portion of its life. The same is true for *Cedarosaurus* and *Astrophocaudia*. It is too bad we don't know more about what was taking place that day in Texas 110 million years ago, when those footprints were made.

8.11 Skeleton of a probable Acro prey, the herbivore *Tenontosaurus*, from Texas. This skeleton stands almost 5 feet (1.5 m) at the shoulders. On display at the Perot Museum of Nature and Science.

Besides dining on young sauropods, Acro may have also preyed on *Tenontosaurus*, a four-legged plant eater about 21 feet (6.4 m) long. This was the other dinosaur found by Stovall and Langston in 1940. How *Tenontosaurus* may have protected itself from attack is a puzzle because it was far smaller than Acro and had no defensive body armor or spikes (fig. 8.11). This tells me that *Tenontosaurus* might have been a major food source unless it had some other defense that doesn't fossilize, such as emitting a horrible smell. We know it was an herbivore because its teeth have flat surfaces for crushing plants (fig. 8.12). An even more vulnerable dinosaur was an as-yet-unnamed, small bipedal dinosaur known as the Proctor Lake hypsilophodontid (hip-sill-oaf-oh-don-tid). At 6–9 feet (1.8–2.7 m) long, this small dinosaur also had no defense except possibly agility in changing directions faster than an Acro and darting into thick underbrush (figs. 8.13 and 8.14). Both *Tenontosaurus* and the Proctor Lake hypsilophodontid may have formed herds (groups? flocks? gaggles? pods? schools?) because several individuals have been found in relatively small areas (*Tenontosaurus*) or together as a group of jumbled skeletons (some of the Proctor Lake hypsilophodontid specimens). Perhaps group protection by herding was how these defenseless herbivores avoided being the main course, a strategy similar to that used by antelopes in Africa today.

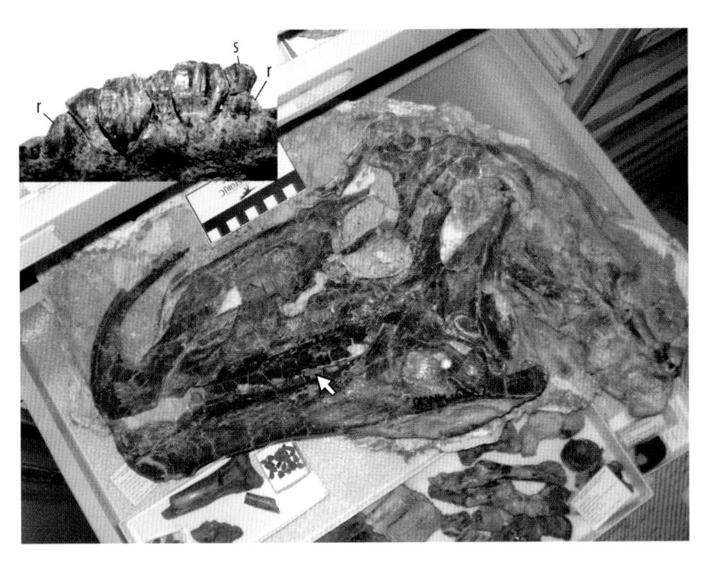

8.12 Skull of *Tenontosaurus* collected in Atoka County, Oklahoma. We know this was an herbivore because the teeth have broad crushing surfaces (arrow), rather than the serrated blades for slicing meat seen in Acro. The broad crushing surface is caused by chewing leaves with grit and twigs. Eventually the teeth get so worn that they are shed (inset is a close-up showing a tooth that was about to be shed, labeled s). Shedding occurs as a replacement tooth (inset shows replacement teeth, labeled r) is pushed up from below. Unlike mammals, which have only two sets of teeth (milk teeth and permanent), dinosaurs pretty much have an unlimited number of replacements.

8.13 Skeleton of another possible prey, an as-yet-unnamed hypsilophodontid from Texas. This small dinosaur, about 18 inches (45 cm) at the shoulder, might have been prey for young Acros.

8.14 An Acro inviting the small hypsilophodontids to lunch. The trees are *Frenelopsis*, which had juniper-like foliage (see fig. 9.5). Grass had not yet appeared, so the dominant groundcover was a variety of drought-resistant ferns. See chapter 9 for further discussion of plants and the environment.

If Acro was the lion of the Early Cretaceous, then *Deinonychus* (dine-on-ick-us) was the jackal. Rich Cifelli and his students from the Sam Noble Oklahoma Museum of Natural History collected a partial skeleton (fig. 8.15) and numerous shed teeth of this theropod near the same place where Langston and Stovall collected the first two *Acrocanthosaurus* specimens (Brinkma, Cifelli, and Czaplewski 1998). Three *Tenontosaurus* skeletons were also found, and some of their bones had grooves that match the teeth of *Deinonychus*. These grooves were made by the teeth of *Deinonychus* while it was feeding on the carcasses of *Tenontosaurus*. Sometimes the teeth would strike the bone and break off, producing the shed teeth found at the site An encounter between an adult Acro and a pack of *Deinonychus* that were feeding on a kill must have been similar to a lion showing up where jackals are feeding. The *Deinonychus* probably surrendered their kill unless they were extremely hungry, such as during a drought, when prey were scarce. Even then, they would have been reluctant to fight the much-larger predator. Maybe they ran around, nipping at the Acro as it scavenged the kill, hoping to drive it away—but I doubt it worked.

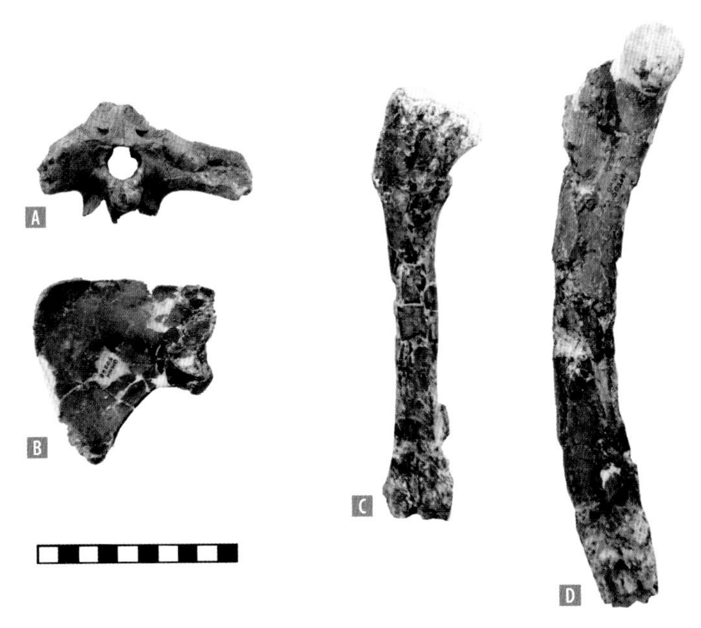

8.15 Some of the handful of bones of a young *Deinonychus* collected by the Sam Noble Oklahoma Museum of Natural History. The material includes **A** the back of the skull, **B** the chest plate, or coracoid, **C** the upper arm bone, or humerus, and **D** the thigh bone, or femur.

CHAPTER 9

The World of Acro

Sedimentary rocks are imprinted by the environment in which they were deposited. For example, the red rock layers, or strata, of the American Southwest indicate that the original sediments were deposited in a hot, dry climate, whereas coal indicates deposition in a hot, wet one (Totman-Parrish, Ziegler, and Scotese 1982). The sedimentary rocks where the Texas and Oklahoma *Acrocanthosaurus* specimens were found indicate a relatively warm and dry climate. By looking at the big picture, we see a world very different from our own.

The Antlers Formation, from which the bones of Acro were collected, was named by Robert Hill (1891) based on exposures of the strata near Antlers, Pushmataha County, Oklahoma. Hill was a prominent Texas geologist during the late nineteenth century and the early part of the twentieth century. He became the leading expert on Cretaceous sedimentary rock layers in the area.

The Antlers Formation is also known by a variety of other names, including the Antlers Sandstone, Trinity Sandstone, and Paluxy Sandstone. Some of these names originated before geologists understood how the various rock layers over large areas are related to one another. An important tool for resolving this issue of the relationships of the rock layers—especially in east Texas and Oklahoma, where plant-free outcrops of rock are rare—was the drill rig, which showed geologists the rock layers below ground. Geologists could then trace the different rock layers underground, and this showed that some of the formations, which in different places had acquired different names, were one and the same. Drilling also showed that the Antlers Formation is up to 984 feet (300 m) thick (Hobday, Woodruff, and McBride 1981). Deep drilling has revealed that the Antlers Formation was deposited on deformed and contorted marine rocks of the Paleozoic age (fig. 9.1). These rocks are exposed today at the surface in the Ouachita and Arbuckle Mountains but continue underground southward into Texas. The youngest of these Paleozoic rocks is the 312-million-year-old Atoka Formation (early Middle Pennsylvanian Period to geologists), meaning there is a gap of more than 200 million years between the Antlers Formation and the underlying Paleozoic rocks. The gap marks a long interval of time of mountain building followed by erosion. First, the ancestral Ouachita and Arbuckle Mountains, along with the Appalachian Mountains, were pushed upward thousands of feet by the slow-motion collision (taking millions of years) that merged the continents into the supercontinent Pangaea (completed about 290 million years ago). Flat-laying sedimentary rocks in southern Oklahoma and western Arkansas were distorted by this collision, as still can be seen from space, like a crumpled rug. Once the mountains stopped rising, they and their surrounding countryside for hundreds of miles underwent about 170 million years of erosion by fast-flowing rivers whose headwaters were in the mountains. By the time Acro appeared, all that remained of the deep gorges that had been cut into the towering mountains were shallow valleys separated by low ridges. Because the rivers no longer flowed rapidly down mountain slopes, the remnants of the mountains were buried under their own debris of sand, silt, and gravel, which became the Antlers Formation.

Pangaea began to break apart about 200 million years ago, forming two separate supercontinents, the northern Laurasia and southern Gondwana. The separation of the North American portion of Laurasia from Gondwana around 170 million years ago opened a waterway that became today's Gulf of Mexico. By 110 million years ago, an arm of the Arctic Ocean, which had begun to creep south across North America 130 million years ago, reached the Kansas-Oklahoma border (fig. 9.2). At this time the ancestral Gulf of Mexico also expanded, moving northward across eastern Texas so that Oklahoma was an isthmus, which I'll call the Isthmus of Oklahoma. This isthmus connected a larger, exposed land to the west with one to the east (fig. 9.3), much like the Isthmus of Panama connects the landmasses of Mexico and South America. Within another 20 million years the two oceans would connect, submerging western Oklahoma and dividing North America in half. This Western Interior Seaway, as it is called by geologists, lasted until a few million years before the end of the Age of Dinosaurs, 66 million years ago.

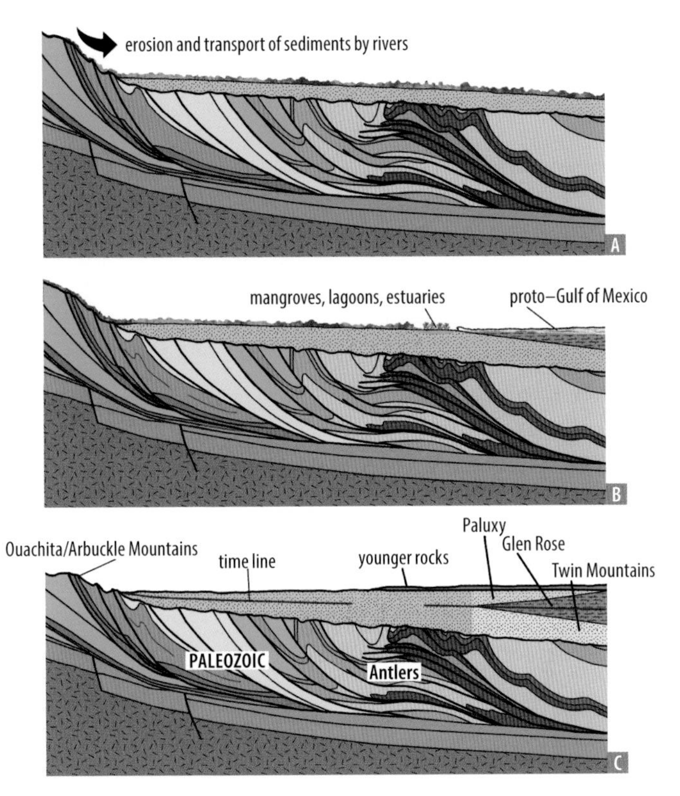

9.1 North-south vertical section through the earth showing the history of the Antlers Formation and related formations. **A** The folded and faulted Paleozoic rocks (actually various shades of gray, green, and red but shown here as various shades of blue and purple) were originally deposited flat on the bottom of the ocean, the proto–Gulf of Mexico. Beginning around 323 million years ago, the southern and eastern part of the North American continent, called a plate, collided with the European plate. For the next 20 million years or so, the two continents were in a slow-motion collision that crumpled the older rocks into the Ouachita-Arbuckle-Wichita mountains. **B** By 115 million years ago, rivers flowed south toward the proto–Gulf of Mexico across a wide, flat coastal plain. Sediments carried by these rivers were deposited on the plain. These sediments (mostly sands and gravels) were to become the lower part of the Antlers Formation in Oklahoma and the Twin Mountains Formation of Texas. Around 113 million years ago, the proto–Gulf of Mexico began to move northward, drowning part of the landscape. Sediments that became the Glen Rose Formation were deposited. Around 110 million years ago, the proto–Gulf of Mexico retreated south, allowing rivers to deposit sands and gravels once more. These became the upper part of the Antlers Formation in Oklahoma and the Paluxy Formation in Texas. Younger sediments would eventually be deposited and largely eroded over the next 105 million years. **C** An imaginary time line, which represents the ground surface at that moment, helps to show how everything is related: while it was dry land in Oklahoma, at that moment the proto–Gulf of Mexico was lapping at the coast in northern Texas. The relationships among the various formations is discussed by Hobday, Woodruff, and McBride (1981).

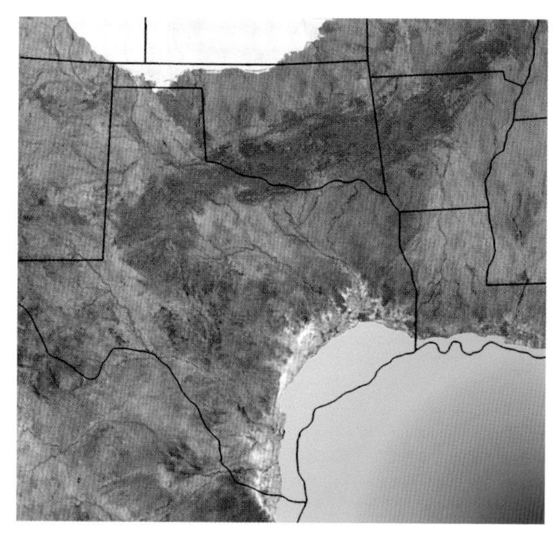

9.2 An imaginary satellite view showing the Ouachita-Arbuckle-Wichita mountain range around 115 million years ago stretching from modern Arkansas to northern Texas (dark band). The proto–Gulf of Mexico extended across part of what is now southern Texas and Louisiana. Estuaries, lagoons, and mangrove swamps occurred along the coast.

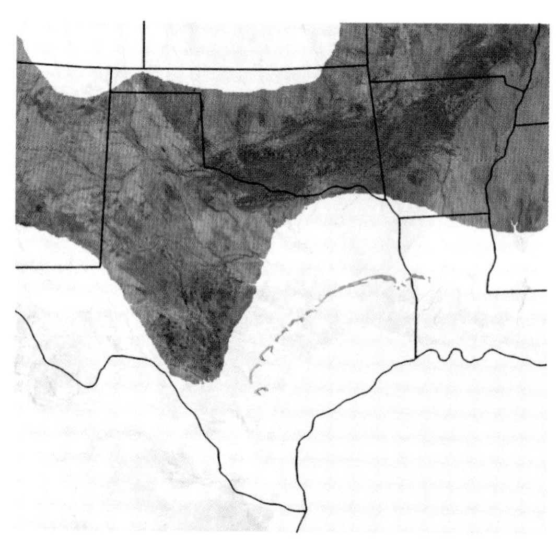

9.3 An imaginary satellite view showing the rise in sea level around 113 million years ago. This sea-level rise was global and brought ocean water down from the proto–Arctic Ocean across central Canada and the United States to northern Oklahoma. A narrow isthmus across Oklahoma connected the eastern and western United States. The shallow and warm proto–Gulf of Mexico was the ideal setting for a barrier reef (irregular blue line) to form across southern Texas. Farther north, the available landscape for the dinosaurs was greatly restricted between the proto–Gulf of Mexico and the mountain range. It is possible that Acro and other dinosaurs were squeezed out of the area at this time.

Our Acro story begins before the submergence of the Isthmus of Oklahoma. At that time, 110 million years ago, southeastern Oklahoma was located about 300 miles (480 km) farther south, at about 29.5° north latitude, rather than the 34° north position where it sits today. Geologists have determined this from measuring the remnant magnetism, or paleomagnetism, in rocks that bear iron minerals. As they were deposited, the magnetic minerals aligned themselves with magnetic north like little compass needles; there, they were locked into place when the rock solidified or formed around them. These magnetic minerals not only point to where magnetic north was 110 million years ago but also, because the earth is a sphere, point at an angle through the earth to where the magnetic pole was at that time. Subtracting this angle (called magnetic inclination) from 90°, which is the North Pole (the equator is 0°), gives the latitude when formation of the rocks "froze" the minerals in place.

Geologists can also infer the climate from sedimentary rock types (Totman-Parrish, Ziegler, and Scotese 1982) as well as from chemical traces in minerals found in ancient soils called paleosols. One common method compares the ratio of isotopes of oxygen, which is the same technique discussed in chapter 6 that Christine Missell used to determine Acro metabolism. As a reminder, there is "light" oxygen, ^{16}O (from adding the number of protons and neutrons) and "heavy" oxygen, ^{18}O, which has two more neutrons. Hydrogen, which combines with oxygen to form water, is not picky about whether it binds with ^{16}O or ^{18}O, but the effects are important for paleoclimatologists, who study ancient climates.

Water made of lighter oxygen evaporates a little more easily than water made of heavier oxygen does; similarly, water vapor with heavy oxygen falls out as rain sooner than vapor with light oxygen does. The relative amounts, or ratios, of these two oxygen water types depend on the temperature. The warmer the temperature (or climate) is, the more light-oxygen water evaporates, leaving behind a greater amount of heavy-oxygen water. In the ocean, this heavier oxygen shows up in the shells of various organisms. On land, this oxygen is found in whitish calcium carbonate lumps or powder in the soil; when the soil becomes rock, these lumps appear as hard nodules. In the sixty years or so since the discovery that the oxygen isotope ratios correspond to temperature, geologists have both refined the technique of measuring the ratios and have built up a large database. Paleoclimatologists now use the oxygen isotope ratios as an ancient thermometer (Dworkin, Nordt, and Atchley 2005).

Rocks and fossils from Oklahoma and Texas that were deposited when Acro was alive show that the average yearly temperature on the Isthmus of Oklahoma was 90°F (32°C). It was most likely in the high 90s to low 100s Fahrenheit (30s Celsius) throughout most of the summer months (Hay and Floegel 2012). This high temperature on the isthmus was due to the high temperatures globally during this time (Totman-Parrish, Ziegler, and Scotese 1982). In fact, it was so warm that there was no ice at the poles (Herman and Spicer 2010). These higher global temperatures lasted for almost 20 million years and mark an interval of time called the Mid-Cretaceous Hothouse. This global warmth was probably caused by high levels of carbon dioxide in the atmosphere from massive volcanic eruptions associated with what has been called the Mid-Cretaceous Superplume Event (Caldeira and Rampino 1991). This superplume was a huge blob of molten rock thousands of miles in diameter that slowly rose toward the surface of the earth. There have been several of these events in the geologic past, with the one during the mid-Cretaceous (roughly 115–90 million years ago) being the most recent. This superplume rose in the South Pacific (roughly between Hawaii and Australia), bulging the crust of the earth upward. This bulge pushed the ocean waters onto the continents, which is why the Isthmus of Oklahoma formed in the first place.

We can follow the progression of sea rise through time across Texas toward Oklahoma (Hobday, Woodruff, and McBride 1981). Around 115 million years ago a vast plain, called the Wichita Paleoplain, stretched to the coastline, which was roughly 25 miles (40 km) inland from where it is today. Across this plain, rivers meandered toward the south and southeast from the rolling hills of the Ouachita, Arbuckle, and other remnant mountains (fig. 9.2). As the proto–Gulf of Mexico slowly rose, its waters crept over the land at a rate of about 3 inches (7.6 cm) per year; this might not seem like much, but it would certainly be noticeable over a lifetime. By about 110 million years ago, the coastline reached what is now southeastern Oklahoma (fig. 9.3). Far out at sea, in an arc extending across central Texas, was a reef formed by odd, extinct clams called rudists, rather than by corals and sponges, which make up reefs today. Rudist shells are shaped like ice cream sugar cones but with little lids on top (fig. 9.4). In some places they grew in such abundance and were packed so close together that little else could grow. Landward from the reefs, the water was very shallow and evaporation was very high in the high temperatures. In this setting, limey muds were deposited on the shallow seafloor, which today forms the rock layer known as the Glen Rose Formation. Nearer the shore, limey mud flats were exposed during low tides, and these

9.4 Reconstruction of part of a rudist clam reef. Although these look like coral, they are in fact clams as shown by the small, lid-like upper shell. These clams were collected in Texas and are now displayed at the Denver Museum of Nature and Science.

became a highway for dinosaurs passing through the region. These are the tracks seen at Dinosaur Valley State Park near Glenn Rose, Texas.

Some really cool studies on various shallow-marine clams living today show that the growth layers in their shells correspond to the seasons and to high-tide–low-tide cycles (e.g., Pannella 1976). These growth layers can be matched to the lengths of lunar months (number of days to go through a complete cycle from full moon to full moon) and lengths of years. Applied to fossil clams, these growth layers show that the number of days in both the lunar month and the year has decreased over geologic time. One of the most detailed studies, by invertebrate paleontologist Giorgio Pannella (1972), shows that when Acro was alive, a day was about 23 hours long, a lunar month was about 30.75 days long, and a year was about 369 days long. Why this difference from today? Because the earth was a little closer to the sun and the moon was a little closer to the earth. Since their formation, the earth has slowly spiraled away from the sun and the moon away from the earth. Acro would have seen a slightly brighter sun and a larger, brighter moon.

Acrocanthosaurus appears to have gone extinct shortly after the proto–Gulf of Mexico reached the Texas-Oklahoma border because no traces of its bones or footprints have been found in the upper part of the Glen Rose Formation and overlying Paluxy Formation. This portion of strata is what geologists refer to as the regressive phase of the Glen Rose, meaning the sea was retreating from the land. As the sea retreated south, the seafloor once again became mudflats, where we would at least expect to find footprints. Nor have Acro bones been found in the river and floodplain deposits of the Paluxy Formation overlying the Glen Rose Formation. It is possible that when the sea was at the Oklahoma-Texas border, the environment was too constricted in the narrow zone just tens of miles wide between the sea and the Ouachita Mountains. With less land to hunt on or feed from, many of the larger dinosaurs were squeezed out of this part of Oklahoma. It is possible that *Acrocanthosaurus* migrated elsewhere, such as what is now northern Wyoming, where the specimen was found in the slightly younger Cloverly Formation (D'Emic, Melstrom, and Eddy 2012).

While Fran was alive, the landscape was very different from what we see today. To reconstruct that landscape, we need to know what fossil plants (wood, leaves, and pollen) have been found in the rocks in which Fran and the other Acro specimens were found. Burnt wood, or fossil charcoal (called fusain), was recovered by paleobotanist Bonnie Jacobs from the rocks at the site where Fran was collected. This shows that wildfires occasionally swept through the area. Such fires were probably caused by lightning strikes. Fossil leaves and wood have also been found in the Antlers Formation, but they have not yet been studied. Instead, we have to draw from the descriptions of fossils from the time-equivalent Twin Mountains and Glen Rose Formations (Fontain 1893; Jacobs 1989).

Plants with flowers, called angiosperms, first appeared around this time. Compared with today, however, they were relatively rare. The landscape was dominated mostly by conifers—cone-bearing trees that today include junipers, pines, cedars, and so forth—of an extinct type called cheirolepidiaceans (kear-oh-lep-id-ace-e-ans). The most common cheirolepideacean was *Frenelopsis* (fren-el-op-sis), known from logs almost 4 feet (1.2 m) in diameter, which suggest the trees were at least 70 feet (21.3 m) tall (Axsmith and Jacobs 2005). Its needle-like leaves looked a lot like those of a juniper (fig. 9.5). In modern plants, needle-like leaves limit water loss in dry environments, so we assume the same must have been true in the ancient past. Because competition for groundwater was high, *Frenelopis* trees were probably widely spaced, forming

9.5 *Left:* Example of a fossil plant, *Frenelopsis*, from the Twin Mountains Formation of central Texas. *Right: Frenelopsis* was a tree with juniper-like foliage to reduce water loss, as can be seen in this reconstruction. This adaptation suggests an arid environment rather than a hot jungle. On display at the Perot Museum of Nature and Science, Dallas, Texas.

an open forest with the ground exposed to the sky rather than a dense forest that blocked sunlight from the ground. Another tree was *Pseudofrenelopsis* (sue-dough-fren-el-op-sis), which had clusters of green, segmented needles.

The groundcover in the bottomlands near the rivers may have included some early angiosperms. They probably grew in shaded areas where the soil was exposed, such as riverbanks below trees (Feild et al. 2004). Their seeds were small so that they could lodge in cracks and crevasses in the soil. Angiosperms also had the advantage of growing roots from shoots that were broken off, such as by a sloppy herbivore feeding or a passing dinosaur trampling the plant. Many of the angiosperms at this time were only a few feet tall, although a few reached the size of a shrub. These shrubs had woody stems for support. Among the fossil angiosperm leaves are two early magnolia-like types (Ball 1937), *Sapindopsis* (sa-pin-dop-sis) and *Araliaephyllum* (ah-ral-ee-fi-lum; fig. 9.6).

Various species of ferns grew in the more open areas away from the river, along with a scattering of shorter conifers or bushes, such as *Glenrosa* (glen-rose-ah; Watson and Fisher 1984), which had scaly, juniper-like needles, and plants called bennettitaleans (ben-net-tea-tail-ee-ans). Bennettitaleans, also known as cycadeoids, superficially look like a cycad (also known as a sago palm). Distinguishing between the two is rather esoteric and requires the use of a microscope to see the pores on the underside of the leaves, but this works only when the leaves are well fossilized. These pores, called stomata, are what plants use to breathe. The pores of bennettitaleans are unique in shape and structure, making them easily identifiable.

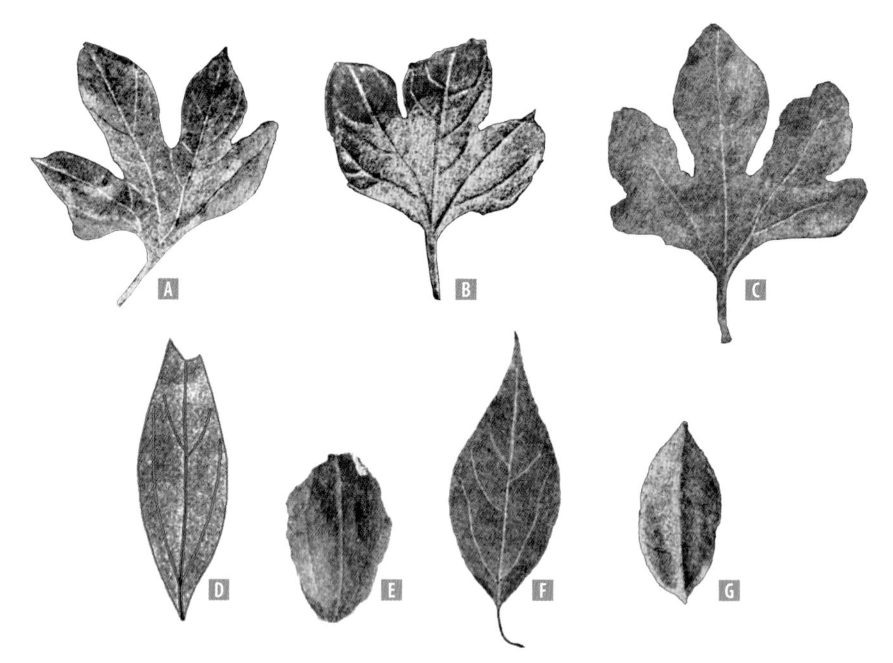

9.6 Flowering plants from the time of Acro are known by their leaves from the Paluxy Formation in Texas. These include leaves from the shrubs or small trees called A B C *Araliaephyllum* and D E F *Sapindopsis.* G Some leaves remain unidentifiable because their preservation is so poor. *Modified and colorized from Ball (1937).*

Dinosaurs are not the only animals found in the Antlers Formation of Oklahoma and the Twin Mountains and Glen Rose Formations of Texas. Some of the earliest work on these nondinosaur fossils was done in 1949 by researchers from the Field Museum of Natural History in Chicago. This work was followed over the next few decades by the University of Texas and Southern Methodist University (Thurmond 1974). In Oklahoma, Rich Cifelli and his students from the University of Oklahoma found hundreds of bones from a variety of animals that help paint a more complete picture of the ecosystem in which Acro lived (Cifelli et al. 1997). Many of these bones in Oklahoma and earlier in Texas were recovered by soaking chunks of soft rock in tubs of water until they crumbled, then washing the crumbled rock through screens, leaving behind a coarse residue. It is in this residue of mostly gravel that the fossils were found. Most of the bones found this way are typically less than 0.5 inches (1.25 cm) long and are called vertebrate microfossils. These bones are important for providing evidence of the many small animals (generally less than 10 pounds, or 4.5 kg, live weight) that lived under the feet of Acro and other dinosaurs. This technique of fossil collecting works best when small bone fragments are found eroding out of the ground, hinting at a concentration beneath the surface.

A variety of aquatic and terrestrial (land-based) animals lived in Acro's world. In the rivers swam at least two types of freshwater shark, *Hybodus* (hi-bode-us) and *Lissodus* (lie-so-dus). These small sharks are known mostly from their small, blunt teeth, which are best suited for crushing aquatic snails, clams, crayfish, and so forth. *Hybodus* may also have eaten small fishes if it managed to catch them. Swimming in the same waters were the gar-like lepisosteids (lee-piz-ah-stee-ids) and the related semionotids (sem-ee-oh-no-tids), which are characterized by diamond-shaped, thick scales covered with enamel (like your teeth) on the outer surfaces. A related fish today, the gar, is a predator that hunts fish and crayfish. Other fishes with crushing teeth were *Coelodus* (see-low-dus), *Gyronchus* (jeer-on-cuss), *Palaeobalistum* (pal-ee-oh-bale-is-tum), and *Proscincetes* (pro-skin-see-tees). These were deep-bodied fish like bluegills, though they are more closely related to the modern bowfin. They had peculiar, small, domed teeth fused to a broad, curved plate of bone; these were probably used to crush aquatic snails, crayfish, and aquatic insects. Another distant relative of the bowfin was the caturid (cat-yur-id) *Macrepistius* (mac-ree-piss-tee-us). Although *Macrepistius* had domed teeth, it and other caturids had a more slender shape, like that of a trout or sardine, unlike *Coelodus* and the other deep-bodied fishes.

Amphibians were represented by the salamander-like *Albanerpeton* (al-ban-urp-ee-ton) and possibly one other genus, but that specimen is too fragmentary to identify. There was also at least one frog, possibly more, but none has been identified as to genus and species. Evidence of aquatic turtles is represented by shell fragments of *Naomichelys* (nah-om-ee-key-lees) and possibly *Glyptops* (glip-tops). Another aquatic turtle, *Trinitichelys* (trin-iti-key-lees), is known from a slender skull and shell (Gaffney 1972). The roof of its mouth has broad crushing surfaces, suggesting it may have competed with the shell-crushing fishes for aquatic snails and possibly crayfish. Lizards are represented by numerous isolated vertebrae and skull pieces. From these, Randall Nydam, who was a graduate student at the University of Oklahoma, and Rich Cifelli (2002) described a new skink called *Atokasaurus* (ah-toke-ah-sore-us) and a whiptail called *Ptilotodon* (till-lot-oh-don; fig. 9.7A). Other bones may belong to an extinct lizard group called the paramacellodids (para-mah-sill-oh-dids), which even then were relics from the Jurassic Period.

Crocodilians are represented by teeth, skull fragments, and the distinctive, pitted-bone armor plates that were embedded in the skin of their necks and backs; some crocs even had a cluster of armor plates on their bellies. Some of these bones belong to a 2-foot-long (61-cm) crocodile called *Bernissartia* (burn-is-sar-tee-ah), first named for a skeleton found in a Belgian coal mine in the early 1880s. Some of the skull fragments and teeth from Oklahoma and Texas may belong to a group of small, short-faced terrestrial crocodiles called atoposaurids (ah-top-oh-saur-ids). These peculiar crocs had hardly any armor in their skin. They also had rather long legs, leading some paleontologists to suggest they may have been fully terrestrial rather than aquatic, as modern crocs are. The largest crocs also belong to an ancient, extinct group called the goniopholids (gone-ee-ah-foal-lids). These crocs looked a lot like the Nile croc, except that their body armor plates had a peculiar peg-and-socket joint between them.

Also scampering around in the underbrush were a variety of shrew- and rat-sized mammals belonging to several primitive lineages (Cifelli 1997; Davis and Cifelli 2011). One of them, called *Astroconodon* (astro-con-oh-don), had peculiar teeth with three pointed cusps aligned in a row. Such piecing teeth suggest a diet of insects; although no insects have yet been found in the Antlers Formation, we can infer they must have been present. Another primitive mammal was the multituberculate (mull-tee-too-burr-cue-lat)

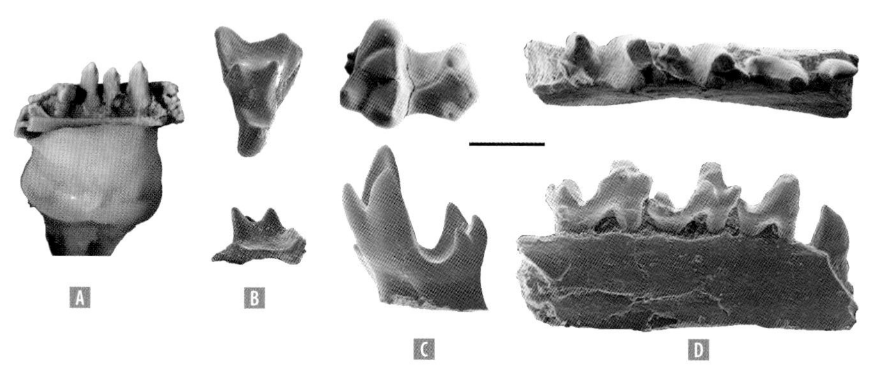

9.7 Tiny jaw fragments and teeth collected by screening soft rock in water and then sorting through the residue. A This lizard jaw fragment belongs to the whiptail *Ptilotodon wilsoni*. It is held to the head of a pin with a small mass of wax. The mammal teeth in top and side views are from B the marsupials *Atokatherium* and C *Oklatheridium* and D the partial jaw of *Slaughteria*, a small, shrew-like mammal. The heavy line is a scale bar 1 millimeter (1/25 inch) long, showing how tiny these specimens are. *Panel A courtesy of Randall Nydam; panels B–D courtesy of Richard Cifelli.*

Paracimexomys (pair-ah-sim-ex-oh-mes). Multituberculates are an extinct branch of early mammals with flat molar teeth covered with numerous crescent-shaped tubercles (cusps), hence the name. They had rodent-like incisors, and many of them had a pair of curved, bladed premolar teeth in the lower jaws for slicing. Multituberculates like *Paracimexomys* are thought to have eaten both plants and insects, meaning they were omnivores. Fragments of other teeth suggest that other species of multituberculates were present, but the material is too fragmentary to identify the genus.

Several early relatives to opossums, or marsupials, are also known from teeth (fig. 9.7), including *Atokatheridium* (atoka-ther-id-ee-um), *Oklatheridium* (oak-lah-ther-id-ee-um), and possibly *Pappotherium* (pap-poh-ther-ee-um). The cheek teeth, or molars, resemble those of various opossums living today, and based on that information, the diet of the fossil opossums can be inferred. The largest of these marsupials, *Oklatheridium*, may have had a varied diet, including lizards, smaller mammals, carrion, insects, and fruits. The smaller opossums, on the other hand, may have fed mostly on insects, which are less likely to put up much resistance. (Can you imagine the indignity of being beaten up by a grasshopper?) Not much is known about the limbs of these early opossums, which is unfortunate because we don't know whether they lived primarily on the ground, in trees, or, like their modern cousins, in both places. The overall small, squirrel-to-mouse size of these Early Cretaceous opossums would certainly not have prevented them from climbing.

Shrew-like mammals are also known from teeth, and these are named mostly for famous mammal paleontologists. These include *Holoclemensia* (ho-low-klem-en-sea-ah), named for Bill Clemens; *Kermackia* (ker-mack-ee-ah), named for Kenneth Kermack, and *Slaughteria* (slaw-ter-ee-ah), named for Bob Slaughter. The tall cusps of these mammals' teeth were ideal for puncturing the hard, outer skeleton of insects, just like shrews do today (fig. 9.7).

From the plants and animals listed in this chapter, you can see that the world of Acro was a mixture of the strange and the familiar. The small, long-legged terrestrial crocs were unlike anything living today, whereas the opossum-like mammal probably would not make you turn your head.

CHAPTER 10

A Final Few Words

In this final chapter, I'll summarize what we know or assume about Acro as well as what we don't yet know. We know that Acro was a *T. rex*-sized dinosaur that lived in Texas, Oklahoma, and Wyoming between 110 and 115 million years ago. The long, thin, bladed teeth are most suited for piercing and cutting flesh, not for pulverizing plant material, so that is how paleontologists know it was a meat-eater. I suspect that it was primarily a predator, hunting and killing rather than feeding on carcasses like a scavenger, because its arms indicate a powerful, muscular grip that was well suited for holding a struggling prey. This interpretation is supported by the large, raptor-like talons of its hand claws. All of this raises the question, what prey did Acro hunt?

There are several possibilities, but there is no consensus among paleontologists. One possibility is one of the plant-eating ornithischians, either *Tenontosaurus* or the smaller, unnamed hypsilophodontid (figs. 8.11 and 8.14). Yet another possibility is suggested by a trackway from a three-toed dinosaur that is thought to have been Acro, based on print size and shape. This trackway seems to follow that of a very large sauropod dinosaur, of which there are four possible candidates: *Sauroposeidon*, *Paluxysaurus*, *Cedarosaurus*, and *Astrophocaudia*. The idea that this Acro was stalking the sauropod comes from the observation that as the sauropod trackway curves toward the left, so does the Acro trackway. Unfortunately, the end of the story remains unknown because the trackways pass into the riverbank, where they remain hidden. Over the years paleontologists have fantasized that a sauropod skeleton would be found at the end of the trackway, but we don't know how much

10.1 *Acrocanthosaurus* walking in a conifer forest during the rainy season, when the understory of ferns was green and rivers flowed. An early flowering plant, *Araliaephyllum*, appears behind the log.

rock overlying the tracks, called the overburden, would have to be removed to uncover the end of the trail. Is it feet, or is it perhaps hundreds or thousands of feet? Of course, it's possible that the ancient landscape contained some barrier that is no longer visible, such as the edge of a lagoon that deflected both the sauropod and Acro, in which case the parallel trackways are incidental.

We do know that lagoons were in the vicinity because of the rocks in which the tracks were preserved (fig. 8.6). Rocks are often a reflection of the environment in which they are deposited, and those of the Paluxy River valley were deposited in very shallow, coastal marine environments (figs. 9.2 and 9.3). Various small- to medium-size rivers flowed south toward the sea, and along the way they occasionally buried dinosaur carcasses and bones (fig. 3.1). These rivers also buried logs, branches, and leaves shed by conifers and the earliest flowering plants. Plants today are very good indicators of climate because of their water and temperature needs, and the same appears to be true for the past as well. The fossil plants from the land of Acro have been studied from only a few sites in Texas, and these indicate a rather arid environment. Acro didn't stomp around in an Amazon-like rainforest. Its environment was much more open, with widely scattered trees (fig. 8.14), except in the low-lying areas near rivers and lakes, where taller conifers may have dominated (fig. 10.1). In the understory near the rivers were some of the earliest flowering plants, the angiosperms, which were bushes and small trees. These rather inconspicuous plants would eventually become the dominant vegetation that we see today.

I do not want to close this chapter leaving you with the impression that we know everything about Acro and its world. In fact, there is much we do not yet know, such the sounds that Acro made when it was threatening a rival or when it was courting. (Was it a growl? A cluck? A chirp?) Nor do we know what color it was, although we may be getting close. There was once a time when I thought dinosaur skin color would forever remain a mystery. But new methods of analysis have revealed coloration patterns as well as traces of skin pigments (Zhang et al. 2010; Wogelius et al. 2011). In a few years, we might have a good idea about what color Acro was—but we'll probably still be wondering if male and female Acros were different colors. Just as exciting is the discovery of soft tissue recovered from within theropod dinosaur bone, where it was sealed off from bacteria and the environment (e.g., Schweitzer 2011). Research on such specimens is still in its infancy, but we may eventually know something about what Acro's blood might have been like. Who knows what the next Acro discovery will reveal? I can hardly wait!

Annotated References

Axsmith, Brian J., and Bonnie Fine Jacobs. 2005. "The Conifer *Frenelopsis ramosissima* (Cheirolepidiaceae) in the Lower Cretaceous of Texas: Systematic, Biogeographical, and Paleoecological Implications." *International Journal of Plant Sciences* 166 (2): 327–37. [Description of a fossil tree material from the Jones Ranch sauropod site, Texas.]

Bailey, Jack B. 1997. "Neural Spine Elongation in Dinosaurs: Sailback or Buffalo-Back?" *Journal of Paleontology* 71: 1124–46. [Discusses the purpose of the tall spines in various dinosaurs and advocates a buffalo-like hump of muscle and fat.]

Ball, Oscar M. 1937. "A Dicotyledonous Florule from the Trinity Group of Texas." *Journal of Geology* 45: 528–37. [Description of sycamore-like plants from the Lower Cretaceous of Texas.]

Barrick, Reese E., and William J. Showers. 1999. "Thermophysiology and Biology of *Giganotosaurus*: Comparison with *Tyrannosaurus*." *Palaeontologia Electronica* 2 (2), http://palaeo-electronica.org/1999_2/gigan/issue2_99.htm. [Examines the oxygen isotope evidence showing that *Gigantosaurus*, a relative of *Acrocanthosaurus*, had an intermediate metabolism between mammals and crocodiles.]

Bates, Karl T. 2009. "Predicting Speed, Gait and Metabolic Cost of Locomotion in the Large Predatory Dinosaur *Acrocanthosaurus* Using Evolutionary Robotics." *Journal of Vertebrate Paleontology* 29 (3, supplement): 59A. [Preliminary results of computer modeling of Acro running speeds.]

Bates, Karl T., Phillip L. Manning, Lee Margetts, and William I. Sellers. 2010. "Sensitivity Analysis in Evolutionary Robotic Simulations of Bipedal Dinosaur Running." *Journal of Vertebrate Paleontology* 30: 458–66. [Results of computer modeling of dinosaur running speeds.]

Behrensmeyer, Anna K. 1978. "Taphonomic and ecologic information from bone weathering." *Paleobiology* 4: 150–62. [A classic study of what happens to a bone exposed to the environment.]

Briggs, Derek E. G. 2003. "The Role of Decay and Mineralization in the Preservation of Soft-Bodied Fossils." *Annual Review of Earth and Planetary Sciences* 31 (1): 275–301. [Summary of the role of microbes in the mineralization of soft tissue, including experimental evidence.]

Brinkman, Daniel L., Richard L. Cifelli, and Nicholas J. Czaplewski. 1998. "First Occurrence of *Deinonychus antirrhopus* (Dinosauria: Theropoda) from the Antlers Formation (Lower Cretaceous: Aptian-Albian) of Oklahoma." *Oklahoma Geological Survey Bulletin* 146: 1–27. [First discovery in Oklahoma of the sickle-footed "raptor" *Deinonychus*.]

Brusatte, Stephen L., and Paul C. Sereno. 2008. "Phylogeny of Allosauroidea (Dinosauria: Theropoda): Comparative Analysis and Resolution." *Journal of Systematic Palaeontology* 6: 155–82. [A thorough review of the allosauroids, which includes the family Carcharodontosauridae and *Allosaurus*, as well as other related theropods.]

Bybee, Paul J., Andrew H. Lee, and Ellen-Thérèse Lamm, 2006. "Sizing the Jurassic Theropod Dinosaur *Allosaurus*: Assessing Growth Strategy and Evolution of Ontogenetic Scaling of Limbs." *Journal of Morphology* 267 (3): 347–59. [Use of lines of arrested growth to age a population of *Allosaurus*, a relative of Acro.]

Caldeira, Ken, and Michael R. Rampino. 1991. "The Mid-Cretaceous Super Plume, Carbon Dioxide, and Global Warming." *Geophysical Research Letters* 18 (6): 987–90. [A peak in global temperatures around the time of Acro correlates with a rise of a giant blob of magma in the Pacific Ocean.]

Carpenter, Kenneth. 1999. *Eggs, Nests, and Baby Dinosaurs: A Look at Dinosaur Reproduction*. Bloomington: Indiana University Press. [Reproductive behavior in dinosaurs and their eggs.]

———. 2002. "Forelimb Biomechanics of Nonavian Theropod Dinosaurs in Predation." *Senckenbergiana Lethaea* 82 (1): 59–75. [Analysis showing the limited range of motion of the arms of carnivorous dinosaurs.]

———. 2005. "Experimental Investigation of the Role of Bacteria in Bone Fossilization." *Neues Jahrbuch für Geologie und Paläontologie Monatshefte* 2005 (3): 83–94. [A detailed description of the experiment using bacteria to preliminarily fossilize modern bone.]

———. 2013. "A Closer Look at the Hypothesis of Scavenging versus Predation by *Tyrannosaurus rex*." In *Tyrannosaurid Paleobiology*, edited by J. Michael Parrish, Ralph E. Molnar, Phillip J. Currie, and Eva B. Koppelhus, 264–77. Bloomington: Indiana University Press. [The olfactory lobes of predators and prey dinosaurs are equally large.]

Carpenter, Kenneth, Dale Russell, Donald Baird, and Robert Denton. 1997. "Redescription of the Holotype of *Dryptosaurus aquilunguis* (Dinosauria: Theropoda) from the Upper Cretaceous of New Jersey." *Journal of Vertebrate Paleontology* 17 (3): 561–73. [Scientific redescription of the first carnivorous dinosaur found in North America.]

Cherel, Yves, Henri Weimerskirch, and Colette Trouvé. 2000. "Food and Feeding Ecology of the Neritic-Slope Forager Black-Browed Albatross and Its Relationships with Commercial Fisheries in Kerguelen Waters." *Marine Ecology Progress Series* 207: 183–99.

Cifelli, Richard L. 1997. "First Notice on Mesozoic Mammals from Oklahoma." *Oklahoma Geology Notes* 57: 4–17. [Describes the fossil mammals found in the Antlers Formation of Oklahoma.]

Cifelli, Richard L., James D. Gardner, Randall L. Nydam, and Daniel L. Brinkman. 1997. "Additions to the Vertebrate Fauna of the Antlers Formation (Lower Cretaceous), Southeastern Oklahoma." *Oklahoma Geology Notes* 57 (4): 124–31. [Gives a list of the nondinosaur vertebrate fossils (fishes, lizards, etc.) collected from the Antlers Formation of Oklahoma.]

Cleland, Timothy P., Michael K. Stoskopf, and Mary H. Schweitzer. 2011. "Histological, Chemical, and Morphological Reexamination of the 'Heart' of a Small Late Cretaceous *Thescelosaurus*." *Naturwissenschaften* 98 (3): 203–211. [The dinosaur "heart" described by Fisher et al. (2000) is shown to be a naturally occurring concretion.]

Crile, George, and Daniel P. Quiring. 1940. "A Record of the Body Weight and Certain Organ and Gland Weights of 3690 Animals." *Ohio Journal of Science* 40 (5): 219–59. [A classical study that records the weights of various organs in a variety of animals.]

Currie, Philip J., and Kenneth Carpenter. 2000. "A New Specimen of *Acrocanthosaurus atokensis* (Theropoda, Dinosauria) from the Lower Cretaceous Antlers Formation (Lower Cretaceous, Aptian) of Oklahoma, USA." *Geodiversitas* 22: 207–46. [Description of the *Acrocanthosaurus* skeleton excavated by Cephis Love and Sid Hall.]

D'Emic, Michael D. 2012. "Revision of the Sauropod Dinosaurs of the Lower Cretaceous Trinity Group, Southern USA, with the Description of a New Genus." *Journal of Systematic Palaeontology* 11 (6): 707–26. [Identifies the long-necked sauropod contemporaries of Acro.]

D'Emic, Michael D., Keegan M. Melstrom, and Drew R. Eddy. 2012. "Paleobiology and Geographic Range of the Large-Bodied Cretaceous Theropod Dinosaur *Acrocanthosaurus atokensis*." *Palaeogeography, Palaeoclimatology, Palaeoecology* 333–34: 13–23. [Briefly describes the bones of a possible immature *Acrocanthosaurus*.]

Dal Sasso, Cristiano, and Simone Maganuco. 2011. "*Scipionyx samniticus* (Theropoda: Compsognathidae) from the Lower Cretaceous of Italy: Osteology, Ontogenetic Assessment, Phylogeny, Soft Tissue Anatomy, Taphonomy and Palaeobiology." *Memorie Società Italiana di Scienze Naturali e Museo Civico di Storia Naturale* 37: 1–281. [A detailed description of a remarkable specimen that includes mineralized soft tissue.]

Daniel, Joseph C., and Karen Chin. 2010. "The Role of Bacterially Mediated Precipitation in the Permineralization of Bone." *PALAIOS* 25: 507–516. [Another experiment using bacteria to fossilize bone.]

Davis, Brian M., and Richard L. Cifelli. 2011. "Reappraisal of the Tribosphenidan Mammals from the Trinity Group (Aptian–Albian) of Texas and Oklahoma." *Acta Palaeontologica Polonica* 56 (3): 441–62. [Describes some of the mammal teeth found by the Sam Noble Oklahoma Museum of Natural History.]

Dworkin, S. I., L. Nordt, and S. Atchley. 2005. "Determining Terrestrial Paleotemperatures Using the Oxygen Isotopic Composition of Pedogenic Carbonate." *Earth and Planetary Science Letters* 237 (1): 56–68. [Using oxygen isotopes of certain rocks to determine temperatures in the geologic past.]

Eddy, Drew R., and Julia A. Clarke. 2011. "New Information on the Cranial Anatomy of *Acrocanthosaurus atokensis* and Its Implications for the Phylogeny of Allosauroidea (Dinosauria: Theropoda)." PLOS ONE 6 (3): e17932. doi:10.1371/journal.pone.0017932. http://www.plosone.org/article/info%3Adoi%2F10.1371%2Fjournal.pone.0017932#pone-0017932-g055. [CT of the braincase of Fran and the approximate shape of its brain.]

Erickson, Gregory M. 1996. "Incremental Lines of von Ebner in Dinosaurs and the Assessment of Tooth Replacement Rates Using Growth Line Counts." *Proceedings of the National Academy of Sciences* 93 (25): 14623–27. [Use of daily growth rings in teeth to infer tooth development and replacement rates in dinosaurs.]

Erickson, Gregory M., A. Kristopher Lappin, and Kent A. Vliet. 2003. "The Ontogeny of Bite-Force Performance in American Alligator (*Alligator mississippiensis*)." *Journal of Zoology, London* 260: 317–27. [Gives the maximum measured bite force in the alligator.]

Farlow, James O. 1981. "Estimates of Dinosaur Speeds from a New Trackway Site in Texas." *Nature* 294:747–48. [Estimates the speed of *Acrocanthosaurus* based on the spacing of footprints.]

———. 1987. *A Guide to Lower Cretaceous Dinosaur Footprints and Tracksites of the Paluxy River Valley, Somervell County, Texas.* Waco, Tex.: Baylor University. [*Acrocanthosaurus* and prey tracks exposed along the Paluxy River, Texas.]

———. 2001. "*Acrocanthosaurus* and the Maker of Comanchean Large-Theropod Footprints." In *Mesozoic Vertebrate Life*, edited by Darren Tanke and Kenneth Carpenter, 408–27. Bloomington: Indiana University Press. [A thorough discussion of why the large three-toed tracks at Dinosaur State Park were made by *Acrocanothosaurus*.]

Farlow, James O., and Eric R. Pianka. 2002. "Body Size Overlap, Habitat Partitioning and Living Space Requirements of Terrestrial Vertebrate Predators: Implications for the Paleoecology of Large Theropod Dinosaurs." *Historical Biology* 16 (1): 21–40. [An analysis of the amount of territory theropod dinosaurs would need if they had a "cold-blooded" or "hot-blooded" metabolism.]

Farmer, C. G., and Kent Sanders. 2010. "Unidirectional Airflow in the Lungs of Alligators." *Science* 327: 338–40. [First announcement that airflow in crocodile lungs was one-way.]

Feild, Taylor S., Nan Crystal Arens, James A. Doyle, Todd E. Dawson, and Michael J. Donoghue. 2004. "Dark and Disturbed: A New Image of Early Angiosperm

Ecology." *Paleobiology* 30 (1): 82–107. [The earliest flowering plants were small plants that grew in shaded areas.]

Ferrell, Russell. 2011. Acrocanthosaurus: *The Bones of Contention.* Ann Arbor, Mich.: Malloy. [Tells the convoluted history of the discovery and subsequent melodrama of Fran, although this version of the story contains some unreliable information.]

Fiffer, Steve. 2001. Tyrannosaurus *Sue: The Extraordinary Saga of Largest, Most Fought over T. rex Ever Found.* New York: Macmillan. [A description of the legal battle over the *Tyrannosaurus* specimen known as Sue.]

Fisher, Paul E., Dale A. Russell, Michael K. Stoskopf, Reese E. Barrick, Michael Hammer, and Andrew A. Kuzmitz. 2000. "Cardiovascular Evidence for an Intermediate or Higher Metabolic Rate in an Ornithischian Dinosaur." *Science* 288 (5465): 503–505. [A large mass in the chest cavity of a dinosaur is interpreted as a fossilized heart.]

Fontaine, William Morris. 1893. "Notes on Some Fossil Plants from the Trinity Division of the Comanche Series of Texas." *Proceedings of the United States National Museum* 16: 261–82. [Early description of the Lower Cretaceous fossil plants from Texas.]

Franzosa, Jonathan, and Timothy Rowe. 2005. "Cranial Endocast of the Cretaceous Theropod Dinosaur *Acrocanthosaurus atokensis*." *Journal of Vertebrate Paleontology* 25 (4): 859–64. [Use of CT scans to study the inside of the braincase of *Acrocanthosaurus* and to reconstruct the approximate shape of the brain.]

Gaffney, Eugene S. 1972. "The Systematics of the North American Family Baenidae (Reptilia, Cryptodira)." *Bulletin of the American Museum of Natural History* 147 (5): 241–320. [Describes the only known specimen of the turtle *Trinitichelyes*.]

Gray, Henry, and Warren H. Lewis. 1918. *Anatomy of the Human Body*. Philadelphia: Lea and Febiger. [This out-of-copyright anatomy book has good illustrations of the human skeleton and muscles.]

Hall, Margaret I., E. Christopher Kirk, Jason M. Kamilar, and Matthew T. Carrano. 2011. "Comment on 'Nocturnality in Dinosaurs Inferred from Scleral Ring and Orbit Morphology.'" *Science* 334: 1641b. [A challenge to the conclusion reached by Schmitz and Motani (2011) that theropod dinosaurs were active mostly at night.]

Hammes, Frederik, and Willy Verstraete. 2002. "Key Roles of pH and Calcium Metabolism in Microbial Carbonate Precipitation." *Reviews in Environmental Science and Biotechnology* 1 (1): 3–7. [Reviews how microbes can precipitate the mineral calcium carbonate from its environment.]

Harris, Jerald D. 1997. "A Reanalysis of *Acrocanthosaurus atokensis*, Its Phylogenetic Status, and Paleobiogeographic Implications, Based on a New Specimen from Texas." Master's thesis, Southern Methodist University. [Jerry Harris's master's thesis describing the Hobson Ranch specimen of Acro.]

———. 1998. "A Reanalysis of *Acrocanthosaurus atokensis*, Its Phylogenetic Status, and Paleobiogeographic Implications, Based on a New Specimen from Texas." *New Mexico Museum of Natural History and Science Bulletin* 13: 1–75. [Published version of Jerry's thesis.]

Hay, William W., and Sascha Floegel. 2012. "New Thoughts about the Cretaceous Climate and Oceans." *Earth Science Reviews* 115 (4): 262–72. [Examines the global climate during the Cretaceous.]

Heard-Booth, Amber N., and E. Christopher Kirk. 2012. "The Influence of Maximum Running Speed on Eye Size: A Test of Leuckart's Law in Mammals." *Anatomical Record* 295: 1053–62. [Faster-moving mammals have larger eyes than their slower-moving close relatives.]

Herman, Alexei B., and Robert A. Spicer. 2010. "Mid-Cretaceous Floras and Climate of the Russian High Arctic (Novosibirsk Islands, Northern Yakutiya)." *Palaeogeography, Palaeoclimatology, Palaeoecology* 295 (3): 409–22. [Fossil plant record of the Arctic region of Siberia shows that the North Pole was ice-free.]

Hill, Robert T. 1891. "The Comanche Series of the Texas-Arkansas Region." *Geological Society of America* 2: 503–24. [Naming of the Antlers Formation in southern Oklahoma.]

Hobday, David K., C. M. Woodruff, and Mary W. McBride. 1981. "Paleotopographic and Structural Controls on Non-marine Sedimentation of the Lower Cretaceous Antlers Formation and Correlatives, North Texas and Southeastern Oklahoma." In "Recent and Ancient Nonmarine Depositional Environments: Models for Exploration," edited by Frank G. Ethridge and Romeo M. Flores. *SEPM Special Publication* 31: 71–87. [Geological history of the Texas and Oklahoma when the sediments of the Antlers Formation were deposited.]

Horner, John R. 2000. "Dinosaur Reproduction and Parenting." *Annual Review of Earth and Planetary Sciences* 28 (1): 19–45. [Summary of Horner's research on parental care in dinosaurs.]

Horner, John R., and Don Lessem. 1994. *The Complete T. rex*. New York: Simon & Schuster. [A case is made for *Tyrannosaurus* being a scavenger rather than predator.]

Jacobs, Bonnie Fine. 1989. "Paleobotany of the Lower Cretaceous Trinity Group, Texas." In *Field Guide to the Vertebrate Paleontology of the Trinity Group, Lower Cretaceous of Central Texas, Society for Vertebrate Paleontology Field Trip Guidebook*, edited by Dale A. Winkler, Phillip A. Murray, and Louis L. Jacobs, 31–33. Dallas, Tex.: Institute for the Study of Earth and Man, Southern Methodist University. [Preliminary discussion of the fossil plants from the Lower Cretaceous of Texas.]

Langston, Wann, Jr. 1947. "A New Genus and Species of Cretaceous Theropod Dinosaur from the Trinity of Atoka County, Oklahoma." Master's thesis, University of Oklahoma. [Langston's master's thesis describing the original two Acro specimens from the Arnold and Cochran farms.]

———. 1974. "Nonmammalian Comanchean Tetrapods." *Geoscience and Man* 8: 77–102. [Summary of the fossil vertebrates from the Lower Cretaceous of Texas, exclusive of fossil mammals.]

Larson, Peter, and Kristin Donnan. 2004. *Rex Appeal: The Amazing Story of Sue, the Dinosaur that Changed Science, the Law, and My Life*. Montpelier, Vt.: Invisible Cities. [Peter Larson's account regarding the legal battle around the *Tyrannosaurus* Sue.]

Lipka, Thomas R. 1998. "The Affinities of the Enigmatic Theropods of the Arundel Clay Facies (Aptian), Potomac Formation, Atlantic Coastal Plain of Maryland." In "Lower and Middle Cretaceous Terrestrial Ecosystem," edited by Spencer G. Lucas, James I. Kirkland, and J. W. Estep. *New Mexico Museum of Natural History and Science Bulletin* 14: 229–34.

Lipkin, Christine, and Kenneth Carpenter. 2008. "Looking Again at the Forelimb of *Tyrannosaurus rex*." In *Tyrannosaurus rex: The Tyrant King*, edited by Peter Larson and Kenneth Carpenter, 166–90. Bloomington, Ind.: University Press. [Restores the arm muscles of *Tyrannosaurus*.]

Lull, Richard Swann. 1953. "Triassic Life of the Connecticut Valley." *State of Connecticut Geological and Natural History Survey* 81: 1–336. [Classic study of the dinosaur footprints of the Connecticut Valley, including a description of a squatting dinosaur.]

Meng, Qingjin, Jinyuan Liu, David J. Varricchio, Timothy Huang, and Chunling Gao. 2004. "Palaeontology: Parental Care in an Ornithischian Dinosaur." *Nature* 431 (7005): 145–46. [The association of an adult skull of *Psittacosaurus* with a mass of juveniles is cited as proof of parental care in this dinosaur.]

Milner, Andrew R. C., Jerry D. Harris, Martin G. Lockley, James I. Kirkland, and Neffra A. Matthews. 2009. "Bird-like Anatomy, Posture, and Behavior Revealed by an Early Jurassic Theropod Dinosaur Resting Trace." *PLOS ONE* 4(3): e4591. Accessed May 4, 2014, http://www.plosone.org/article/info%3Adoi%2F10.1371%2Fjournal.pone.0004591#pone-0004591-g007.

Miser, Hugh D. 1954. *Geologic Map of Oklahoma*. Norman, Okla.: U.S. Geological Survey and Oklahoma Geological Survey. [Map (scale 1:500,000) used to identify the rock formation where Fran was found.]

Missell, Christine Ann. 2004. "Thermoregulatory adaptations of *Acrocanthosaurus atokensis*—Evidence from oxygen isotopes." Master's thesis, North Carolina State University, Raleigh. [A study using oxygen isotopes to determine that the metabolism of Fran was more bird- and mammal-like than alligator-like.]

Motani, Ryosuke, Bruce M. Rothschild, and William Wahl. 1999. "Large Eyeballs in Diving Ichthyosaurs." *Nature* 402: 747. [Shows that largest eyeballs occurred in *Temnodontosaurus* for seeing in low light in deep water.]

Nydam, Randall, and Richard Cifelli. 2002. "Lizards from the Lower Cretaceous (Aptian–Albian) Antlers and Cloverly Formations." *Journal of Vertebrate Paleontology* 22: 286–98. [Describes the fossil lizards from the Antlers Formation.]

O'Connor, Patrick M. 2006. "Postcranial Pneumaticity: An Evaluation of Soft-Tissue Influences on the Postcranial Skeleton and the Reconstruction of Pulmonary Anatomy in Archosaurs." *Journal of Morphology* 267: 1199–1226. [Examines the air bladder system in a variety of birds and asserts that traces left on vertebrae of theropods provide evidence that they had a similar breathing system.]

O'Connor, Patrick M., and Leon P. A. M. Claessens. 2005. "Basic Avian Pulmonary Design and Flow-through Ventilation in Non-avian Theropod Dinosaurs." *Nature* 436: 253–56. [Discusses the probable presence of bird-like pouch outgrowth from the breathing system throughout the body of theropods, especially *Majungatholus*.]

Ortega, Francisco, Fernando Escaso, and José L. Sanz. 2010. "A Bizarre, Humped Carcharodontosauria (Theropoda) from the Lower Cretaceous of Spain." *Nature* 467 (7312): 203–206. [Description of *Concavenator*, which had a "sail" composed of two vertebrae.]

Osborn, Henry F. 1905. "*Tyrannosaurus* and Other Cretaceous Carnivorous Dinosaurs." *Bulletin of the American Museum of Natural History* 21: 259–65. [Naming of *Tyrannosaurus* and a brief description of the bones.]

Pannella, Giorgio. 1972. "Paleontological Evidence on the Earth's Rotational History Since the Early Precambrian." *Astrophysics and Space Science* 16: 212–37. [Use of growth rings in various fossils reveals the change in the length of day through time as the earth spirals away from the sun.]

———. 1976. "Tidal Growth Patterns in Recent and Fossil Mollusc Bivalve Shells: A Tool for the Reconstruction of Paleotides." *Naturwissenschaften* 63 (12): 539–43. [Ancient tidal cycles revealed by growth layers in fossil shells.]

Rauhut, Oliver W. M. 2011. "Theropod Dinosaurs from the Late Jurassic of Tendaguru (Tanzania). *Special Papers in Palaeontology* 86: 195–239. [Reviews the carnivorous dinosaurs collected from the famous Tendaguru site, in what is now Tanzania, and names *Veterupristisaurus*.]

Rayfield, Emily J. 2005. "Using Finite-Element Analysis to Investigate Suture Morphology: A Case Study Using Large Carnivorous Dinosaurs." *Anatomical Record Part A: Discoveries in Molecular, Cellular, and Evolutionary Biology* 283 (2): 349–65. [Analysis of the stresses generated in an *Allosaurus* skull during a bite.]

Reid, Robert E. H. 1993. "Apparent Zonation and Slowed Late Growth in a Small Cretaceous Theropod." *Modern Geology* 18: 391–406. [Destruction of the innermost lines of arrested growth as an individual theropod grows.]

———. 1997. "How Dinosaurs Grew." In *The Complete Dinosaur*, edited by James O. Farlow and Michael K. Brett-Surman, 403–13. Bloomington: Indiana University Press. [Summary of bone physiology in dinosaurs.]

Romer, Alfred Sherwood, and Llewellyn W. Price. 1940. "Review of the Pelycosauria." *Geological Society of America Special Papers* 28: 1–534. [Description of the sail-backed early relative of mammals and a discussion of the sail to control body temperature.]

Ruben, John A., Cristiano Dal Sasso, Nicholas R. Geist, Willem J. Hillenius, Terry D. Jones, and Marco Signore. 1999. "Pulmonary Function and Metabolic Physiology of Theropod Dinosaurs." *Science* 283 (5401): 514–16.

Schachner, Emma R., John R. Hutchinson, and C. G. Farmer. 2013. "Pulmonary Anatomy in the Nile Crocodile and the Evolution of Unidirectional Airflow in Archosauria." *PeerJ* 1: e60. http://dx.doi.org/10.7717/peerj.60. [Online, detailed study of one-way flow of air through the crocodile lung.]

Schmitz, Lars, and Ryosuke Motani. 2011. "Nocturnality in Dinosaurs Inferred from Scleral Ring and Orbit Morphology." *Science* 332: 705–708. [Gives data to estimate eyeball size from the dimension of the sclerotic ring and also concludes that most predatory dinosaurs were night creatures.]

Schweitzer, Mary Higby. 2011. "Soft Tissue Preservation in Terrestrial Mesozoic Vertebrates." *Annual Review of Earth and Planetary Sciences* 39: 187–216. [Summary of soft tissue preservation in fossils, especially dinosaurs.]

Scully, Crispian. 2002. *Oxford Handbook of Applied Dental Sciences.* Oxford: Oxford University Press. [Gives the bite force of humans.]

Senter, Phil, and James Robins. 2005. "Range of Motion in the Forelimb of the Theropod Dinosaur *Acrocanthosaurus atokensis*, and Implications for Predatory Behaviour." *Journal of Zoology*, 266: 307–318. [A study of the range of motion of Acro based on moving around casts of arm bones.]

Sereno, Paul C., Didier B. Dutheil, M. Larochene, Hans C. E. Larsson, Gabrielle H. Lyon, Paul M. Magwene, Christian A. Sidor, David J. Varricchio, and Jeffrey A. Wilson. 1996. "Predatory Dinosaurs from the Sahara and Late Cretaceous Faunal Differentiation." *Science* New Series 272: 986–91. [Brief description of the first complete skull of *Carcharodontosaurus*.]

Shinya, Akiko, and L. Bergwall. 2007. "Pyrite Oxidation: Review and Prevention Practices." Poster presented at the Preparators' Session, 67th Annual Meeting of the Society of Vertebrate Paleontology, Austin, Tex. http://vertpaleo.org/PDFS/0c/0cf8d5c7-d1a1-4a0a-96b2-a28658f9b4cf.pdf. [A poster that goes into detail about what pyrite decay does to fossil bone and how to treat it.]

Stovall, J. Willis, and Wann Langston, Jr. 1950. "*Acrocanthosaurus atokensis*, a New Genus and Species of Lower Cretaceous Theropoda from Oklahoma." *American Midland Naturalist* 43: 696–728. [Published account of Wann Langston's master's thesis.]

Stromer, Ernst. 1931. "Wirbeltier-Reste der Baharije-Stufe (unterstes Cenoman). 10. Ein Skelett-Rest von *Carcharodontosaurus* nov. gen." *Abhandlungen der Bayerischen Akademie der Wissenschaften Mathematisch-naturwissenschaftliche Abteilung* 9: 1–31. [Detailed German description of the carnivorous dinosaur *Carcharodontosaurus*.]

Tanke, Darren H., and Philip J. Currie. 1998."Head-Biting Behavior in Theropod Dinosaurs: Paleopathological Evidence." *Gaia* 15: 167–84. [Documents bite marks on the skull bones of carnivorous dinosaurs.]

Thomas, David A., and James O. Farlow. 1997. "Tracking a Dinosaur Attack." *Scientific American, 277* (6): 74–79. [Describes the Acro stalking and attacking a sauropod.]

Thurmond, John T. 1974. "Lower Vertebrate Faunas of the Trinity Division in North-Central Texas," *Geoscience and Man* 8: 103–29.

Totman-Parrish, Judith, A. M. Ziegler, and Christopher R. Scotese. 1982. "Rainfall Patterns and the Distribution of Coals and Evaporites in the Mesozoic and Cenozoic." *Palaeogeography, Palaeoclimatology, Palaeoecology* 40 (1): 67–101. [Shows the correlation between certain rocks and climate.]

Tracy, C. Richard, J. Scott Turner, and Raymond B. Huey. 1986. "A Biophysical Analysis of Possible Thermoregulatory Adaptations in Sailed Pelycosaurs." In *The Ecology and Biology of Mammal-like Reptiles*, edited by Nicholas Hotton,

Paul D. MacLean, Jan J. Roth, and E. Carol Roth, 195–206. Washington, D.C.: Smithsonian Institution. [A detailed study about body temperature control in sail-backed animals.]

Walls, Gordon L. 1963. *The Vertebrate Eye and Its Adaptive Radiation*. Bloomfield Hills, Mich.: Cranbrook Press. https://archive.org/details/vertebrateeyeits00wall. [A classic book on the eye in all the major vertebrate groups.]

Watson, Joan, and Helen L. Fisher. 1984. "A New Conifer Genus from the Lower Cretaceous Glen Rose Formation, Texas." *Palaeontology* 27: 719–27. [Description of a small conifer adapted for a very dry environment from the Lower Cretaceous of Texas.]

Wedel, Mathew J. 2003. "Vertebral Pneumaticity, Air Sacs, and the Physiology of Sauropod Dinosaurs." *Paleobiology* 29: 243–55. [Presence of air bladders into the vertebrae of sauropod dinosaurs]

Wedel, Mathew J., Richard L. Cifelli, and R. Kent Sanders. 2000. "*Sauroposeidon proteles*, a New Sauropod from the Early Cretaceous of Oklahoma." *Journal of Vertebrate Paleontology* 20 (1): 109–14. [Description of the giant sauropod *Sauroposeidon*, known by a string of neck vertebrae.]

Witmer, Larry M. 1997. "The Extant Phylogenetic Bracket and the Importance of Reconstructing Soft Tissues in Fossils" In *Functional Morphology in Vertebrate Paleontology*, edited by Jeff Thomason, 19–33. Cambridge: Cambridge University Press. [Introduces the concept of extant phylogenetic bracketing as a way to understand what soft tissue might have been in extinct animals.]

Wogelius, Roy A., Phillip L. Manning, H. E. Barden, N. P. Edwards, S. M. Webb, W. I. Sellers, K. G. Taylor, P. L. Larson, P. Dodson, H. You, L. Da-qing, and U. Bergmann. 2011. "Trace Metals as Biomarkers for Eumelanin Pigment in the Fossil Record." *Science* 333 (6049): 1622–26. [Use of synchrotron X-ray on a Cretaceous bird to reveal color patterns in the feathers.]

Zhang, Fucheng, Stuart L. Kearns, Patrick J. Orr, Michael J. Benton, Zhonghe Zhou, Diane Johnson, Xing Xu, and Xiaolin Wang. 2010. "Fossilized Melanosomes and the Colour of Cretaceous Dinosaurs and Birds." *Nature* 463 (7284): 1075–78. [Report of skin pigment traces in a feathered dinosaur from China.]

Zhao, Qi, Michael J. Benton, Xing Xu, and Martin J. Sander. 2014. "Juvenile-Only Clusters and Behaviour of the Early Cretaceous Dinosaur *Psittacosaurus*." *Acta Palaeontologica Polonica* 59: 827–33. [Reanalysis of the adult and baby *Psittacosaurus* association indicates the adult skull was added.]

Index